P9-BAV-237

Folk Medicine

The Art and the Science

Richard P. Steiner, *Editor*
University of Utah

American Chemical Society
WASHINGTON, DC
1986

Library of Congress Cataloging in Publication Data

Folk medicine.
Includes bibliographies and index.

1. Materia medica, Vegetable. 2. Medicinal plants. 3. Folk medicine.

I. Steiner, Richard P., 1946- . [DNLM: 1. Medicine, Herbal. 2 Medicine, Traditional. 3. Plants, Medicinal. QV 766 F666]

RS164.F64 1985 615′.321 85-22904
ISBN 0-8412-0939-1
ISBN 0-8412-0946-4 (pbk.)

PRINTED IN THE UNITED STATES OF AMERICA

About the Editor

Richard P. Steiner is presently executive officer and adjunct associate professor in the chemistry department at the University of Utah. He received his B.S. from the University of Michigan and his Ph.D. from the University of Wisconsin. He is an active member of the American Chemical Society Division of Chemical Education and has organized numerous meetings and symposia. It is in this capacity that he first became involved in the area of folk medicine. Other areas of interest include designing curriculum for elementary school children, teaching chemistry to nonscience students, and developing techniques to enhance the learning of chemistry. These interests have led to 10 papers and numerous invited presentations at schools and meetings.

Contents

Preface

The medical practices of non-Western cultures, both past and present, are shrouded in mystery. Critics discount many of these practices as chicanery, and proponents often embrace everything without the discriminating analysis needed when examining any field. In the constant effort to improve the efficacy of Western medicine, researchers are increasingly turning their attention to folk medicine for new drugs. The purpose of this book is to examine the medical practices of non-Western cultures in order to establish a scientific basis for the successes of these remedies. The chapters are written in a manner designed to satisfy the curiosity of the lay reader with a modest chemistry background as well as to serve as a reference for researchers in the field.

Several approaches can be taken to establish the scientific basis for the successes of folk medicine. One method is to begin with an anthropological approach in order to appreciate how and why a particular medical practice became established and then to explore the chemical and biological reasons for the success of the practice. This approach is taken in the first chapter of this book and is continued in the next two chapters. These chapters also introduce a second approach, that of describing the available plants. In a broad sense this "botanical-survey" (looking for plants with specific effects) approach is the predominant theme in the next four chapters. The last seven chapters use another approach: They are looking for the active ingredient in a specific medicinal plant. These chapters place a greater emphasis on the isolation and identification of the specific chemical components of the plant.

This area of research brings together workers from many countries and explores many different cultures. In an attempt to emphasize this diversity, contributions in this book come from 9 different countries and discuss the medicines of 11 different cultures.

I thank the contributors for their cooperation. I also acknowledge the support of Merck Sharp & Dohme Research

Laboratories, Pfizer Company, and the United Nations Educational, Scientific, and Cultural Organization for the financial support that allowed us to bring many of these people together for the symposium upon which this book is based.

RICHARD P. STEINER
University of Utah
Salt Lake City, UT 84112

July 1985

Chapter 1

Aztec Sources of Some Mexican Folk Medicine

BERNARD ORTIZ DE MONTELLANO*

Much of Mexican folk medicine is a result of the reconciliation of the Aztec medicinal beliefs and European medical theories of the 16th and 17th centuries. Most analyses of Mexican folk medicine *(1)* have emphasized the Spanish heritage as being more rational or scientific, particularly with respect to the "hot-cold" concept of disease, which is supposed to be a simplified version of the Hippocratic doctrine of humors. However, perpetuation of these medicinal practices in folklore often does not depend on apparent scientific rationales but rather on the customs practiced by the people, particularly peasants. These customs tend to be passed from generation to generation, particularly if these beliefs can be reconciled with those of the conquerors (as was the case in Mexico) and if the remedy is effective on empirical grounds. This chapter will illustrate the perpetuation of Aztec folklore and of the use of two particular plants that were closely identified with Tlaloc, the Aztec rain god.

Fundamental to this premise is some evidence that the present culture and beliefs in Mesoamerica (a cultural area that included Mexico, Guatemala, and Honduras) remained intact despite the rise and fall of several empires prior to the arrival of the Spanish (the Spanish Conquest, 16th century). The use of the backstrap loom (the basic instrument of weaving), which is identical in

*Anthropology Department
Wayne State University
Detroit, MI 48202

appearance in a Jaina funerary statuette (A.D. 800), in a post-Conquest Aztec codex [the Mendoza (1541)], and in today's marketplace in Oaxaca (1980), is an example of such evidence. Further evidence is the fact that the huipil, the basic Mayan dress, and the design of Mayan houses have not changed fundamentally in more than 1000 years.

The peasants' cultural beliefs endured as well as their tangible possessions. Before the Conquest, the Aztecs believed that pregnant women should not go out at night in order to avoid seeing a lunar eclipse (*metzqualoniliztli,* "the eating of the moon"). As the name states, the belief was that bites were being taken out of the moon and that if the event was witnessed by a pregnant woman, her baby would be born with a harelip, that is, a bitten lip, presumably by sympathetic magic. To ward off this event, the woman would place an obsidian knife next to her belly *(2)*. Hundreds of years later, in Tecospa, an Aztec village near Mexico City, Madsen *(3)* found that this belief remained in every detail, except for the substitution of a *metal* knife as the protective charm. I have since related this story in places far removed from Mexico or the Aztecs (Salt Lake City, 1972; Denver, 1976; Detroit, 1980) and have found that the belief exists in the Chicano community. The details are identical, except for the substitution of a metal key for the knife as the protective charm.

Many of the customs perpetuated through the ages had their roots in the religious beliefs of the early cultures. Among the extensive Mesoamerican pantheon, Tlaloc, probably more than any other god, held the loyalty of the stolid peasant. As god of rain, lightning, and agricultural fertility, Tlaloc held the key to the success of farming in areas such as the Basin of Mexico, where drought was feared. Remarkable consistency is found in depictions of Tlaloc in pre-Columbian Mesoamerica throughout more than 1000 years. Tlaloc's attributes and iconography survived in a remarkably uniform fashion despite the rise and fall of several "civilizations" and empires in the Valley of Mexico. This survival occurred presumably because the common people were always in residence, with or without elites, nobles, or a hierarchy of priests. The cult of the god that was important to the commoners would be upheld even if the more esoteric gods of the elites were neglected.

CONCEPTS OF DISEASE

In most cultures, three basic beliefs of the causes of illness are present: (1) diseases caused by divine punishment, (2) diseases caused by the ill will of other beings through the vehicle of magic, and (3) diseases due to natural causes. The Aztecs believed in all of these, and in particular associated many diseases with gods. Specific gods were usually associated with particular symptoms and ailments. Tlaloc was particularly associated with diseases caused by water and cold such as edema, rheumatism, leprosy, and gout *(4)*. Table I lists diseases caused by Tlaloc or other members of his group. Death brought about by drowning or by lightning and diseases such as edema were a sign of having been chosen by Tlaloc to live after death in his paradise. Persons who survived lightning or strong winds were afflicted by "fear illnesses". Some diseases were believed to be the consequence of violating rituals associated with feasts dedicated to Tlaloc or those related to the manufacture of pulque (a fermented drink). The patron gods of such feasts were part of Tlaloc's complex. In the case of *coacihuiztli* (gout, paralysis, or stiffness), cold and dampness are specifically cited as part of the disease's etiology, and a number of other symptoms are associated with phlegm in the chest.

The association of two particular plants with this rain god complex and how this association can reveal the complex interactions between the physical and the supernatural world in Aztec belief have been explored *(5, 6)*. The two plants are *Tagetes lucida (yauhtli, pericon;* African day flower) and *Artemisia mexicana (iztauhyatl, estafiate;* wormwood). Table II summarizes the nonmedical ritual uses of *yauhtli* and *iztauhyatl* described in the basic sources about Aztec culture. These herbs are always mentioned in the context of or in relation to gods or ceremonies of the water god complex and these herbs are *not* specifically mentioned in any other contexts. These herbs thus stand in a special relationship to Tlaloc. Once this connection is established, the process can be reversed. By studying the medicinal uses listed for these plants, one can determine the full range of disease states associated with this god complex. Tables III–V list all the ailments for which these herbs are cited. Each table groups diseases that have something in common.

Table I. Ailments Associated with Tlaloc and Water God's Complex

Aztec Disease	*Literal Translation*	*Equivalent*	*Death by Tlaloc*	*Illness Sent or Possession*[a]	*Other Remarks*
Teococoliztli	god's disease	leprosy	X[b]	—	breaking fast in Atamalcualiztli feast[c]
Coacihuiztli	stiffening of the serpent	gout, paralysis, stiffness	X[d]	X[e]	associated with cold and wet,[f] violation of pulque-making ritual[g]
Atemaliztli	swelling due to water	edema, swelling, buboes	X[d]	—	—
Atonahuiztli	aquatic fever	intermittent fever (malaria ?)	—	X[h]	phlegm in chest[i]
Aacqui	lightning struck, suffering intrusion	epilepsy, madness, insanity	—	X[j]	phlegm in chest,[i] possession via lightning[i]
Netonalcahualiztli	abandonment of *tonalli*	loss of *tonalli*[k] by fear	—	—	narrow miss by lightning or strong wind (*susto*)[i]
Mocuitlapammauhtique	she whose back is terrified	*susto* by lightning[l]	—	—	narrow miss by lightning or strong wind (*susto*)[i]
Netlahuitequiliztli	struck by lightning	death by lightning	X[g]	—	also sent by Chalchiutlicue[m]

[a] Illness sent or possession by Tepictoton, Tlaloque, Auaque, and Chaneque.
[b] Dibble, C. E.; Anderson, A. J. O., Eds. and Translators; "Florentine Codex: General History of the Things of New Spain," Books 1–12; Univ. of Utah Press: Salt Lake City, 1950–69; Book 1, p. 68.
[c] de Sahagun, B. "Historia General de las Cosas de la Nueva Espana," 2d ed.; Porrua: Mexico City, 1969; Vol. 1, p. 230.
[d] Dibble, C. E.; Anderson, A. J. O., ibid., Book 3, p. 45.
[e] Ibid., p. 74.
[f] *See* Ref. *6*.
[g] Dibble, C. E.; Anderson, A. J. O., ibid., p. 49.
[h] Lopez Austin, A. "Cuerpo Humano e Ideologia: Las Concepciones de los Antiguos Nahuas"; Univ. Nacional Autonoma de Mexico: Mexico, 1980; Vol. 1, p. 389.
[i] *See* Ref. *7*.
[j] Lopez Austin, A., ibid., p. 407.
[k] Ibid., p. 394; Dibble, C. E.; Anderson, A. J. O., ibid., Book 1, p. 68.
[l] Ibid, p. 176.
[m] Lopez Austin, A., ibid., p. 392.

Table II. *Yauhtli* and *Iztauhyatl* in Water God's Ritual

Month	*Ritual*
Atlcualo[a]	sacrifice to Tlaloque called *yauhqueme* (dressed in *yauhtli*); ceremony on top of mountain by same name
Etzalcualiztli[b]	(1) chief priest uses *yauhtli* as incense; (2) *yauhtli* scattered before Chalchiutlicue; (3) *yauhtli* used as incense and scattered before Tlaloc; (4) spectators at sacrifices carry *iztauhyatl* flowers of *iztauhyatl* used to scare worms from children's eyes
Tecuilhuitontli[c]	(1) feast of Huixtocihuatl (salt goddess and elder sister of Tlaloque); her staff adorned with *yauhtli* flowers; (2) dancers wore *iztauhyatl* garlands; (3) ceremonial sacrifices carried *yauhtli* and *iztauhyatl*
Atemoztli[d]	(1) *yauhtli* burned to invoke rain; (2) Opochtli (a Tlaloque and god of boatmen); during his feast *yauhtli* is offered and *yauhtli* used in ritual

God(dess)	*Ritual*
Tzapotlatenan (goddess in water god group)	requires *yauhtli*[e] incense
Xilonen (corn goddess related to water and fertility)	requires *yauhtli*[f] in feast
Chalchiutlicue	priests strewed[g] *yauhtli* before her
Tlaloc	nicknamed *yauhtli*; worshipped with rubber and *yauhtli* in his heaven Tlalocan[h]
Tzentzon totochtin (400 rabbits-pulque gods-agricultural)	(1) associated with Tlaloc; pulque is cold[i]; (2) Tepoztecatl; one of them is in region of Yauhtepec mountain[j]

[a]Dibble, C. E.; Anderson, A. J. O., Eds. and Translators; "Florentine Codex: General History of the Things of New Spain," Books 1-12; Univ. of Utah Press: Salt Lake City, 1950-69, Book 2, pp. 43-44.
[b]Ibid., pp. 78-90.
[c]Ibid., pp. 92-94.
[d]Ibid., pp. 151-54; Book 1, p. 21.
[e]Ibid., Book 2, p. 208.
[f]Ibid., pp. 106-7.
[g]Ibid., Book 1, p. 21.
[h]Ibid., Book 6, pp. 35-39, 115.
[i]Lopez Austin, A. "Cuerpo Humano e Ideologia: Las Concepciones de los Antiguos Nahuas," Univ. Nacional Autonoma de Mexico: Mexico, 1980; Vol. 1, p. 295.
[j]Dibble, C. E.; Anderson, A. J. O., ibid., Book 1, pp. 45, 47; Nicholson, H. B. In "Handbook of Middle American Indians"; Wauchope, R., Ed.; Univ. of Texas Press: Austin, 1971; Vol. 10, p. 408.

Table III lists diseases causing phlegm in the chest precipitated by aquatic gods. A brief description will be presented here, but fuller evidence is available elsewhere *(7)*. The Aztecs believed that reasoning resided in the heart and that disturbances could produce a variety of physical syndromes, varying from evil behavior to epilepsy and madness. Thus, the presence of phlegm in the chest (apart from producing fever because phlegm was hot) could disturb the heart by pressuring it, by pushing it aside and "crushing" it *(yolpatzmiquiliztli)*, or by "spinning it around like a spindle" *(teyolmalacacholiztli)*. A slow accumulation of phlegm was believed to produce illnesses such as fever *(7)*. However, under frightening circumstances, such as those caused by narrowly missing a lightning bolt or by an apparition, the phlegm was believed to accumulate rapidly, resulting in illnesses caused by pressure and displacement of the heart.

The diseases listed in Table IV seem to have an element of divine origin combined with an element of exposure to cold. There is a clear connection in the case of *teococoliztli*, "god's illness", which is usually translated in dictionaries as leprosy but is probably some other skin disease. As seen in Table I, *coacihuiztli* is regarded as punishment by Tlaloc, but *coacihuiztli* is also believed to be caused by cold winds. A combination of both divine and natural causes is shown in Tables I and IV. The cure involves both medication and appeals to the deity.

Table III. Pre-Columbian Uses of *Iztauhyatl* and *Yauhtli*

Disease Entity	Iztauhyatl	Yauhtli
Aquatic fever (intermittent fever)	X[a]	X[a–e]
Aftereffects of wind or lightning	X[a, d]	X[b–e]
Fevers and excess phlegm	X[a, d]	X[a, c, e]
Madness, epilepsy, ailments from oppression of the heart	X[a, c]	X[a–c]

NOTE: The etiology for all was phlegm in the chest. The cure was rational (expel phlegm).

[a]Dibble, C. E.; Anderson, A. J. O., Eds. and Translators; "Florentine Codex: General History of the Things of New Spain," Books 1–12; Univ. of Utah Press: Salt Lake City, 1950–69; Book 11.

[b]Hernandez, F. "Historia Natural de las Cosas de la Nueva España"; Univ. Nacional Autonoma de Mexico: Mexico, 1959.

[c]Ximenez, F. "Cuatro Libros de la Naturaleza y Virtudes Medicinales de las Plantas y Animales de la Nueva España"; Escuela de Artes: Morelia, 1888.

[d]de la Cruz, M. "Libellibus de Medicinalibus Indorum Herbis"; Instituto Mexicano Seguro Social: Mexico, 1964.

[e]Lopez Austin, A. *Estud. Cultural Nahuatl*, **1971**, *9*, 125.

Table V lists diseases whose etiology seems to be primarily physical: an excess of either water or cold, both expected characteristics of a rain god. In these cases, the rational cure would involve a diuretic to get rid of the excess water and/or a hot medicine to counteract the cold. In fact, *yauhtli* and *iztauhyatl* do

Table IV. Pre-Columbian Uses of *Iztauhyatl* and *Yauhtli*

Disease Entity	Iztauhyatl	Yauhtli	*Etiology*	*Cure*
Gout (*coacihuiztli*), paralysis, stiffness	X[a]	X[a]	divine plus cold	religious plus rational (hot)
Leprosy (*teococoliztli* = "divine illness")	—	X[b]	divine	religious
Relapse from illness	X[a, c, d]	—	divine plus cold	religious
Spitting blood	—	X[a]	?	?

[a]Dibble, C. E.; Anderson, A. J. O., Eds. and Translators; "Florentine Codex: General History of the Things of New Spain," Books 1–12; Univ. of Utah Press: Salt Lake City, 1950–69; Book 11.
[b]Ibid., Book 10.
[c]de la Cruz, M. "Libellibus de Medicinalibus Indorum Herbis"; Instituto Mexicano Seguro Social: Mexico, 1964.
[d]Hernandez, F. "Historia Natural de las Cosas de la Nueva España"; Univ. Nacional Autonoma de Mexico: Mexico, 1959.

Table V. Pre-Columbian Uses of *Iztauhyatl* and *Yauhtli*

Disease Entity	Iztauhyatl	Yauhtli	*Etiology*	*Cure*
Diuretic, cleaning of kidneys	X[a]	X[b, c]	excess water	rational (diuretic)
Swellings and blisters	X[a-d]	X[b, c]	excess water	rational (diuretic)
Relieves thirst of edema	—	X[b, c]	excess water	rational (diuretic)
Pains of cold origin	X[b]	—	cold	rational (hot)
Cough, dry nostrils	X[a, d]	X[b, c]	cold	rational (hot)
Digestive troubles (empacho), children's colic	X[b, c, e]	X[b, c]	cold	rational (hot)

[a]Dibble, C. E.; Anderson, A. J. O., Eds. and Translators; "Florentine Codex: General History of the Things of New Spain," Books 1–12; Univ. of Utah Press: Salt Lake City, 1950–69; Book 11.
[b]Hernandez, F. "Historia Natural de las Cosas de la Nueva España"; Univ. Nacional Autonoma de Mexico: Mexico, 1959.
[c]Ximenez, F. "Cuatro Libros de la Naturaleza y Virtudes Medicinales de las Plantas y Animales de la Nueva España"; Escuela de Artes: Morelia, 1888.
[d]Dibble, C.E.; Anderson, A.J.O., ibid., Book 10.
[e]Lopez Austin, A. *Estud. Cultural Nahuatl,* **1971,** *9,* 125.

contain diuretics and diaphoretics that would produce heat and perspiration *(8, 9)*.

A widespread belief in many cultures, including the Aztec, is that certain diseases are the result of the insertion under the victim's skin of small objects such as pieces of bone, flint, or small pebbles through sorcery or magic. The remedy usually involves suction of the offending object. Aztec healers chewed *iztauhyatl* and sprayed it on the victim's skin before sucking out the object *(10)*. Although object intrusion is no longer considered to be a major source of illness in Mexican or Chicano folk medicine, the use of *iztauhyatl* in a similar fashion has lasted until the present. *Iztauhyatl* is an important component in the ritual cleansings, *limpias,* that are often applied in folk medicine *(11)*. In an example of reconciliation, these *limpias* often include two herbs of European origin: rue *(Ruta graveolens)* and rosemary *(Rosmarinus officinalis)*. These herbs were classified as hot in the Hippocratic system used in Europe, just as *iztauhyatl* and *yauhtli* were in the Aztec system. As will be discussed, these herbs have apparently become the generic equivalents of the Aztec herbs.

COLONIAL PERIOD

Much of the following analysis is based on the seminal work of Aguirre Beltran *(12)*. Aztec religion came under intense pressure after the Conquest. The Catholic Church was engaged in wholesale conversions to Christianity; these conversions also entailed efforts to root out all traces of previous beliefs. The Inquisition during the 13th century was the primary mechanism for the enforcement of purity of belief in the New World. As shown by Aguirre Beltran, the Inquisition's records tell us much about the native religion during the colonial period. At the same time, the Catholic Church encouraged reconciliation of the contending religions in various ways. The Catholic Church used hieroglyphs to teach the catechism to the natives (Testerian writing), and the Catholic Church worshipped deities, such as the Virgin of Guadalupe or Our Lord of Chalma, at sites that had been previously consecrated to the Aztec gods Tonantzin ("our little mother") and Oztoteotl ("the god of the cave") *(13)*. Using Aztec worship sites may have been unintended, although a number of Franciscans warned against these cults at the time *(14)*.

After the Conquest, and as the old Aztec priests died off and were not replaced, isolated native groups had difficulty maintaining the complex Aztec pantheon. In the absence of a formally trained priest, the shaman *(curandero)* became an agent of cultural preservation. The rain god became more important in religious terms. The *curandero* was closely tied to the rain god by a number of characteristics that identified him as "one of the chosen". Two characteristics were that he could regulate hail and rainfall (this attribute is still important) and that at some point he was supposed to have visited Tlalocan, the paradise of the rain god, either in dreams or in psychedelic visions. A strong shamanic element was present, including the use of psychedelics and divination. Before the Conquest, these practices had apparently been more important in the rites of gods other than Tlaloc. As Tlaloc gained importance as the major deity, his herbs, *iztauhyatl* and *yauhtli,* were also used increasingly for divination and generalized curing rites *(15)*.

Before the Conquest, Aztec religion had been characterized by its generous attitude toward the adoption of gods from other areas. A temple in the capital housed the images of the patron gods of conquered tribes. These gods were cared for reverently. This attitude facilitated the reconciliation of elements from Catholicism. The Virgin of Guadalupe or Our Lord of Chalma was probably not clearly perceived as Christian in the minds of the natives. In the domain of medicine, plants associated with rituals were eventually referred to by Christian names so the natives could use them in ceremonies and avoid being denounced to the Inquisition: *peyotl* became "Mary's rose", *ololiuhqui* became "Our Lord", *yauhtli* became "Saint Mary's herb", *atlinan* became "Our Lady of the Waters", etc. *(16)*. Thus, under pressure from the Inquisition, these name changes made it possible to continue old practices and prayers under the guise of Christian healing.

The natives probably also adopted rue and rosemary, two herbs of European origin, as generic substitutes for *iztauhyatl* and *yauhtli.* As we have seen, the latter were considered to be hot and were used against cold diseases in the Aztec system. The former two were classified as hot in the Galenical European system. Evidence for this adoption can be seen in the increasingly overlapping uses of rue and rosemary and of *iztauhyatl* and *yauhtli* in colonial medicine shown in Tables VI–VIII.

Comas *(17)* has described as "reverse acculturation" the process by which New World physicians trained in the classical European theory adopted New World remedies into the pharmacopoeia. From the beginning, a great amount of European interest existed in the medicinal products of the New World. An example is the rapid diffusion of a book by Monardes describing medicinal plants of the Americas. This book was published in 1569, 1571, and 1574 and was translated into English by 1577. Further evidence of reverse acculturation is the large amounts of sarsaparilla, chinchona, and Holy Wood exported by the colonies for medicinal uses. European physicians in the New World adopted a number of Aztec remedies, among them *iztauhyatl* and *yauhtli*. The majority of the uses for these plants were identical with those in pre-Columbian medicine, as can be seen in Table VI. The rationale for their use, however, had been stripped of the Tlaloc connection and rationalized according to Galenical notions of the opposition of hot and cold qualities. Some contemporary observers claim that the presence of a hot-cold theory of disease causation in modern Mexican folk medicine is due solely to Spanish influence and that pre-Columbian medical theory was devoid of the concept *(18)*. Table VI shows that we are dealing with Aztec

Table VI. European Colonial Uses of *Yauhtli* and *Iztauhyatl*

Ailment	Iztauhyatl	Yauhtli	*Rue* (Ruda)	*Rosemary* (Romero)	*Aztec Use*
Epilepsy	X[a, b]	—	—	—	yes
Edema	X[a, c]	X[b]	—	—	yes
Obstructed kidney	X[a, c]	X[b]	—	—	yes
Stomach, vomiting	X[a, b]	X[b]	—	—	yes
Colic	X[a, b]	—	—	—	yes
Madness, craziness	X[c]	X[b]	—	X[c]	yes
Pains of cold origin	X[b, c]	—	—	—	yes
Intermittent fever	—	X[d]	—	—	yes
Worms	X[a, c]	—	—	—	no
Purgative	X[c]	—	—	—	no
Provoke menstruation	X[a]	X[b]	X[c]	—	no
Against witchcraft	—	X[c]	X[c]	—	?

NOTE: ? indicates unknown.
[a]Farfan, A. *American Indigena,* **1954,** *14,* 327.
[b]de Vetancurt, A. *Anales de Antropologia,* **1968,** *5,* 129.
[c]Lopez, G. *Anales de Antropologia,* **1964,** *1,* 145.
[d]Ruiz de Alarcon, H. "Tratado de las Idolatrias, Supersticiones, Dioses, Ritos, Hechicerias y otras Costumbres Gentilicas de las Razas Aborigenes"; Fuente Cultural: Mexico, 1953.

remedies for Aztec ailments, and these uses are precisely those that are preserved in folklore to the present. It is worth noticing that at this early time in the colony, rue and rosemary are only used for one of the traditional Aztec ailments.

Kay *(19)* has argued that the similarity of remedies used in folk medicine by various groups in the Southwest must be due to a common source of transmission. Kay proposed as this fundamental source de Esteyneffer's *Florilegio Medicinal (20)*, first published in 1712 and widely reprinted thereafter and used in missions throughout the Southwest. Table VII lists those ailments in *Florilegio Medicinal* for which *iztauhyatl* and *yauhtli* are prescribed and that correspond to pre-Columbian uses. The increasing generic correspondence between *iztauhyatl-yauhtli* and rue-rosemary can be seen in the much greater overlap in their uses compared to those in the earlier colonial sources of Table VI. Table VIII lists those additional medicinal uses of *iztauhyatl* for which no Aztec precedent exists. Presumably these arose from orthodox Galenic theory. If Foster and Kay are correct and the primary source of folk medicine is European, we should expect *yauhtli* and *iztauhyatl* to be used in modern folk medicine primarily for European uses rather than for uses of pre-

Table VII. Remedies in *Florilegio Medicinal* for Aztec Ailments Cured by *Yauhtli* and *Iztauhyatl*

Ailment	*Estafiate* (Iztauhyatl)	*Rue*	*Rosemary*	*Santa Maria* (Yauhtli)	*Still Used in SW and Mexico*
Intermittent fever	X	X	X	X	X
Perlesia, paralysis	X	X	—	—	X
Gout, rheumatism	X	—	X	—	X
Cold disease	X	X	X	—	X
Fear	X	X	—	X	X
Empacho	X	X	—	—	X
Against sorcery	—	X	X	—	X
Lightning burns	—	X	—	—	X
Colic	—	X	—	—	X
Epilepsy	X	X	X	—	—
Edema	X	X	X	—	—
Phlegm explusion	X	X	—	—	—
Diuretic	—	—	—	X	—
Madness, melancholia	X	—	—	—	—
Fevers	—	X	—	—	—

Table VIII. Non-Aztec Uses of Estafiate, *Ruda* and *Romero* in *Florilegio Medicinal*

Ailment	*Estafiate* (Iztauhyatl)	*Rue*	*Rosemary*	*Still Used in SW and Mexico*
Deafness (pituita)	X	X	X	X
Diarrhea, dysentery	X	—	—	X
Earworms	X	X	—	—
Worms	X	X	—	—
Liver obstruction	X	—	X	—
Wound treatment	—	X	X	—
Pujos (intestinal pain)	X	X	—	—
Hard breasts	X	X	—	—
Flegmon (swelling)	X	X	—	—
Rabies	X	X	—	—
Scurvy	X	—	X	—
Prevention of apoplexy	—	X	X	—
Blindness	—	X	X	—
Pain in womb (from vinegar)	—	X	X	—
Sciatica	—	X	X	—
Syphilis	—	X	X	—
Vomiting	X	—	—	—
Genital inflammation	X	—	—	—
Bad breath	—	X	—	—
Obstructed spleen	X	—	—	—

Columbian origin. The exact opposite is the case, as can be seen by comparing the uses listed in Table IX with those of Tables VII and VIII. Of the 11 surviving uses in the 20th century, 9 were also used in pre-Columbian medicine and only 2 were European introductions. This difference is statistically significant with a χ^2 value corresponding to 0.04.

MODERN PRACTICES

Many folk medicinal and ritual ceremonies cannot be truly understood without knowledge of the special relationship between the rain god and certain illnesses and their cures. Madsen's study of Tecospa *(21)* is a case in point. Madsen found a widespread belief that dwarfs, water spirits that live in caves in the surrounding mountains, are responsible for a number of cold diseases. This belief corresponds exactly to the characteristics of the Tlaloque or Tepictoton listed in Table I. A particular ailment called *yeyecacuatsihuiztli* is believed to be caused by cold air

Table IX. Modern Folk Uses of *Iztauhyatl* and *Yauhtli*

Disease Entity	Iztauhyatl	Yauhtli	*Other Hot Herbs*
Susto (fright)	X[a, b]	X[c]	rue, rosemary
Lightning fright	—	X[d]	—
Witches	X[e]	—	—
Fears of pregnant women	—	X[e, f]	—
Aire de cuevas (cave air)	—	X[g]	rue, rosemary
Yeyecacuatsihuiztli (=*ehecacoacihuiztli*), rheumatism + paralysis	X[g]	—	rue
Rheumatism	X[a, e, g]	—	—
Malaria (=*atonahuiztli*)	—	X[a]	—
Cough	X[e]	X[a]	—
Colic, empacho	X[c, e, h]	X[a, e, i]	—
Stomach ailments	X[a, e, g, j, k]	X[l, m]	—
Diarrhea	X[e]	—	—

[a]Martinez, M. "Las Plantas Medicinales de Mexico"; Ediciones Botas: Mexico, 1959.
[b]Isunza Ogazon, A. In "Estudios Sobre Etnobotanica y Antrophologia Medica"; Viesca Treviño, C., Ed.; IMEPLAM: Mexico, 1976.
[c]Madsen, W. "The Virgin's Children: Life in an Aztec Village Today"; Greenwood: New York, 1969.
[d]Diaz, J. L. In "Estado Actual del Conocimiento en Plantas Medicinales Mexicanas"; Lozoya, X., Ed.; IMEPLAM: Mexico, 1976.
[e]Ford, K. C. "Las Yerbas de la Gente: A Study of Hispano-American Medicinal Plants"; Univ. of Michigan Museum of Anthropology: Ann Arbor, 1975.
[f]Lewis, O. "Life in a Mexican Village: Tepoztlan Revisited"; Univ. of Illinois Press: Urbana, 1963.
[g]Madsen, W. *J. Am. Folklore,* **1955,** *68,* 123.
[h]Alvarez, L. In "Estudios Sobre Etnobotanica y Antrophologia Medica"; Viesca Treviño, C., Ed.; IMEPLAM: Mexico, 1976.
[i]Gonzalez, J. F. "La Flora de Nuevo Leon"; Imprenta Catolica: Nuevo Leon, 1888.
[j]Candelaria, A., Las Cruces, New Mexico, personal communication.
[k]Instituto Medico Nacional, "Datos para la Materia Medica Mexicana," Secretaria de Fomento, 1898.
[l]Mak, C. *Am. Indigena,* **1955,** *19,* 12.
[m]Logan, M. *Antropos,* **1973,** *65,* 537.

emanating from caves, and the symptoms include paralysis and pain in the joints. The recommended treatment includes *iztauhyatl,* rue, and other hot plants. As shown in Table IX, a correct transcription of the name of the disease, *ehecacoacihuiztli* (the wind god's paralysis), confirms the pre-Columbian etiology, as shown in Table I, and the remedy.

Viesca Trevino *(22)* is currently researching a guild of curers called *graniceros* ("hail people"), who have strong shamanic attributes and who are reputed to hold off hail, to attract rain, to protect crops, and to cure people. The group undertakes a yearly pilgrimage on May 3–5 (the days of the Holy Cross) to a cave on the side of a mountain near Mexico City called Ixtaccihuatl in order to renew their powers. Firecrackers are used to simulate lightning in order to attract beneficent beings. The cave is swept out with a broom made of *yauhtli* to dispel two "evil winds" called "the Bull" and "the Snake" because these spirits can cause the *graniceros* to lose all their powers. Inside the cave, a blue cross (the traditional color for Tlaloc in Aztec symbolism) is worshipped. Here again a reconciliatory combination of Aztec and Christian beliefs and the perpetuation of the characteristic of the control of hail are shown. Garibay *(23)* describes the practice of warding off hail and attracting rain with *iztauhyatl* before the Conquest, and the connection between Tlaloc and lightning is shown in Table I. These examples (and others beyond the scope of this chapter) show reconciliation, but in this case magic, religion, and medicine have all played a role in the perpetuation of *yauhtli* and *iztauhyatl* as Tlaloc's herbs.

AZTEC WOUND TREATMENT

Medical practices may also endure because experience validates their use over time. Previous work *(24, 25)* has shown that the Aztecs had empirically developed medicines that produced physiological effects in accordance with their etiological beliefs. One would also expect their intense military activity would yield ample opportunity to practice and experiment with various treatments for wounds. The treatments they developed compared quite favorably with those of the contemporaneous Spanish conquerors. Cortes, who was cured by Indian doctors after being wounded in a battle, wrote to Charles V that it was not necessary to send doctors from Spain because the Indians had many good ones available *(26)*.

The most authentic source of information about Aztec medicine is the information gathered by Bernardino de Sahagun between 1547 and 1585. Bernardino de Sahagun questioned different sets of native informants and recorded the answers in Nahuatl. The examples given here are taken from the *Florentine Codex:*

> Wounding of the nose.... The severing of the nose is thus helped; one's nose which has been cut off, is replaced, sutured with hair. It is bathed with salted bee honey.... *(27)*

> The pulpy maguey leaf is pounded with a stone.... The juice of the small maguey...is boiled in an olla...; salt is mixed in. With it is healed one who is wounded in the head, or someone who is cut somewhere, or whom they have knifed. It is placed there where the head is wounded or where he is cut.... *(28)*

As the previous examples point out, a principal agent in the treatment of wounds was the concentrated sap of the maguey (*Agave* spp.). The following citation *(29)* outlines the recommended treatment in all its steps:

> ...(head wound).... First the blood is quickly washed away with hot urine, and when it has been washed, then hot maguey sap is squeezed thereon. When it has been squeezed out on the place where the head is wounded, then once again maguey sap, to which are added (the herb) called matlalxihuitl and lampblack with salt stirred in, it is placed on it. And when this has been placed on, then it is quickly wrapped in order that air will not enter there, and so it heals. And if one's flesh is inflamed (this medicine) is placed two or three times....

Similar uses of maguey are given elsewhere in Book 10 *(30)* as well as by some of Sahagun's earlier informants *(31)*. Significantly, similar treatments are given in Books 10 and 11 because different sets of informants were involved in the elaboration of the texts. Other early sources that also cite the use of maguey sap in wounds are Hernandez *(32)*, Motolinia *(33)*, and de las Casas *(34)*. This use of maguey persisted through colonial times *(35)* and can still be found *(36)*.

The sequence from Sahagun's best informants is as follows: wash with warm, that is, fresh, urine; treat with *matlalxihuitl (Commelina pallida)* as a styptic hemostat; apply a concentrated sap expressed from maguey leaves (*Agave* spp.) with or without added salt; and allow the natural processes to take over, unless signs of infection appear, requiring repetition of the treatment. The individual steps will be discussed separately.

Urine

This is a rational agent, particularly if the available water is of unknown quality. In normal men and women, urine is sterile *(37)*.

Commelina pallida

This species contains tannins that provide a styptic effect by precipitation of proteins. This species is listed as a hemostat in the Mexican pharmacopeia *(38)*. Although *C. pallida* has not been studied extensively, *C. pallida* might contain potent vasoconstrictors and smooth muscle contractors, because it produces effects that cannot be attributed to tannins. In the late 19th century, researchers at the National Medical Institute in Mexico investigated the physiological effects of this plant. The techniques of the time were somewhat primitive; whole plant extracts, rather than isolated compounds, were used. Nevertheless, the results are intriguing. An extract of *C. pallida* placed on the exposed intact jugular of pigeons caused contraction of the vein until the circulation was cut off. *C. pallida* caused contraction of smooth muscles when extracts were injected intradermally into pregnant dogs and rabbits. The uterine contractions caused these animals to abort *(39)*. These vasoconstrictor effects as well as the ability of extracts to accelerate coagulation of blood were again demonstrated by Perez-Cirera *(40)*. These actions cannot be due solely to tannins but are consistent with the presence of alkaloids of the lysergic acid type. Further chemical investigation of this plant would be rewarding.

Support for the validity of the use of this species comes from the fact that *Commelina* species are also used as hemostats in China *(41)*. This use had to arise independently of Aztec usage.

Maguey Sap

The main chemical components of interest are the polysaccharides [approximately 10% by weight in the dilute sap *(42)*] and saponins. Saponins and sapogenins are widely distributed in *Agave* species. The most common types are gitogen, hecogenin, tigogenin, manogenin, diosgenin, and smilagenin *(43–45)*. Saponins can produce a variety of effects, among them hemolysis and detergency. Their abundant use in folk medicine for skin infections may be explained by the antibiotic and fungistatic activity generally exhibited by saponins *(46)*. Other investigations show that some saponins exhibit antiviral activity, probably due to their effect on surface tension *(47)*. Even if the saponins actually present in the *Agave* species that concern us do not show all of these activities, saponins could make a significant contribution to an antiseptic effect.

Ample precedent exists for assuming that the bactericidal effect of concentrated maguey sap lies in part with the high polysaccharide content. Majno *(48)* has investigated the medicinal properties of honey and grease used by the Egyptians to treat wounds. He found that heat-sterilized honey inoculated with bacteria and kept at room temperature killed both *Staphyloccocus aureus* and *Escherichia coli.* Several reasons why honey should be bactericidal exist. One is that, being hypertonic, the osmotic pressure would desiccate the bacterial cells, killing them in the process. Majno *(49)* reports that honey was used in wounds in Shanghai during World War II and that "despite its own stickiness it prevents the dressing from sticking to the wound because it draws out a large amount of fluid, and this is said to have a cleansing effect, especially useful on dirty or infected wounds...." Thus, this hypertonic effect may play more than one role in the healing of wounds. Not only does honey have a direct bactericidal effect, but also honey promotes a flow of serum to the wound area, encouraging the natural defense mechanisms. This strategy *(50)* was used successfully by the British in World War I, after it was discovered that commonly used antiseptics were more likely to be harmful if used after a wound had become infected:

> Sir Almroth Wright...applied hypertonic solution to the wound; by this means a flow of serum was induced and this fresh serum exerted a stronger antibacterial action on microbes lying in its path....

Recent reports *(51–53)* provide clinical and experimental support for the use of sugar in infected wounds. Herszage et al. *(51)* treated a series of 120 infected wounds by (1) widening the wound opening, (2) drying the tissues with gauze, and (3) filling the wound with as much sugar as possible and adding more sugar periodically. The cure rate was greater than 99% even in the case of difficult wounds. Chirife et al. *(52, 53)* investigated the theoretical basis for this activity. He measured osmotic pressure by defining the water activity of a solution (a_w) as the ratio of its vapor pressure (p) to that of pure water at the same temperature (p_0) so that $a_w = p/p_0$. Every microorganism has a limiting a_w

below which the microorganism will not grow. For human bacterial pathogens such as streptococci, *Klebsiella, E. coli, Clostridium perfringens,* and *Pseudomonas,* the minimum level of a_w is 0.91 or more. *S. aureus,* however, can tolerate a lower a_w. It is able to grow at 0.86. Because of this, Chirife chose this species for extended experiments and found that sucrose solutions of a_w = 0.86 completely inhibited the growth of *S. aureus* in a series of controlled experiments. This provides a scientific basis for the role of sugar. Packed granulated sugar in wounds will gradually dissolve and maintain an environment of sufficiently low water activity to inhibit the growth of bacteria. Bose *(54)* pointed out that honey would be preferable to granulated sugar because honey has greater osmotic pressure and lower pH (3.7 versus 7.0), as well as being more readily available to the poor rural inhabitants of the Third World.

Majno *(55)* reports that dilution of honey up to 50% with physiological saline (thereby reducing the sugar concentration to 40%) did not decrease its bactericidal activity appreciably. Concentrated sap would approximate the same sugar concentration.

Davidson and Ortiz de Montellano *(56)* tested maguey sap for antibacterial activity; the procedure and the results are summarized here. Commercially available maguey syrup was purchased and tested by saturating blank 14-mm antibiotic sensitivity disks with the syrup, placing these disks in the center of Mueller-Hinton plates that had been streaked with bacterial cultures, and measuring the zones of inhibition resulting after 24 h of incubation at 37 °C. This method was based on a previously reported procedure *(57)* and was pretested with a known antiseptic sticky exudate from *Myroxylon balsamum (58)*.

The procedure is as follows. The maguey syrup was prepared in five different forms: (1) undiluted, (2) 20 mL of maguey syrup diluted with 1 mL of sterile distilled water (a 20:1 dilution), (3) 20 mL of maguey syrup diluted with 5 mL of sterile distilled water (a 4:1 dilution), (4) 20 mL of maguey syrup combined with 5 mg of NaCl (a 4.28×10^{-3} molar solution), and (5) 20 mL of maguey syrup combined with 1 mg of NaCl (a 8.56×10^{-4} molar solution). The organisms were selected on the basis of their role in causing or contributing to aerobic wound infections. The results are presented in Table X.

Table X. Zones of Inhibition of Bacteria by Maguey Syrup

	Zone of Inhibition (mm)[a]				
Bacteria	*Undiluted Syrup*	*+1.0 mL of Water*[b]	*+5.0 mL of Water*	*+0.5 mg of Salt*	*+1.0 mg of Salt*
Salmonella paratyphi[c]	50	42	40	46	48
Pseudomonas aeruginosa[c]	45	38	38	40	43
Escherichia coli[c]	38	27	22	33	38
Shigella sonnei[c]	27	25	25	25	25
Sarcina lutea[d]	23	17	16	21	23
Staphylococcus aureus[d]	20	17	17	33	35

[a]The zone of inhibition is measured as the diameter of the circle.
[b]The plus (+) indicates the addition of salt or water to the maguey syrup.
[c]This is a Gram-negative enteric bacteria, rod shaped.
[d]This is a Gram-positive pyogenic cocci bacteria.

The following implications can be drawn from the results illustrated in Table X. The maguey syrup inhibits bacterial growth. There is variation in the effectiveness of the syrup that depends on the physical characteristics of the organisms. Statistical analysis of the differences between the zones of inhibition exhibited by the six organisms indicates a P value of less than 0.0001. The undiluted syrup is effective against all organisms except *S. aureus.* The addition of salt to the syrup does not enhance the antibacterial properties in five of the six organisms tested. The only organisms whose susceptibility to maguey syrup is increased through the addition of salt are *S. aureus.*

DISCUSSION

Because the pyogenic bacteria *(Staphylococcus aureus* and *Sarcina lutea)* are known to play a significant role in skin invasion including superficial infections such as impetigo and deep follicular lesions including sycosis vulgaris, furuncolosis, and carbuncles *(59, 60)*, the observation that salt increases the effectiveness of maguey syrup for the pyogenic bacteria *S. aureus* demonstrates that the Aztec healers probably included salt in their preparation of maguey syrup. The data by Chirife provide a possible explanation for the difference in results with *S. aureus* when salt is added to the syrup. *S. aureus* is the most resistant organism to dehydration due to its low a_w and is therefore least inhibited by the maguey syrup alone. Adding salt increases the osmotic pressure and thus increases its zone of inhibition. In the case of the other microorganisms that have a higher a_w, the syrup alone already exceeds the a_w needed to inhibit growth. Additional salt is superfluous and produces no significant increase in inhibition.

The object of this work has been twofold: (1) to test the effectiveness of an ancient and popular herbal remedy and (2) to contribute toward the compilation of traditional remedies to which beneficial, harmful, and/or neutral properties can be ascribed. This objective is stated by the World Health Organization *(61)* and discussed by Etkins *(62)*. Perhaps continued laboratory analysis of traditional medicinal remedies used extensively in underdeveloped countries will enable health care providers to recommend the continued utilization of traditional remedies in those cases where substantial scientific evidence exists that these products can provide a safe and inexpensive alternative to Western medicine.

LITERATURE CITED

1. Foster, G. M. *J. Am. Folklore,* **1953,** *66*, 201.
2. Dibble, C. E.; Anderson, A. J. O., Eds. and Translators; "Florentine Codex: General History of the Things of New Spain," Books 1–12; Univ. of Utah Press: Salt Lake City, 1950–69; Book 7, p. 8.
3. Madsen, W. "The Virgin's Children: Life in an Aztec Village Today"; Greenwood: New York, 1969; p. 75.
4. Dibble, C. E., Anderson, A. J. O., ibid., Book 1, p. 47.
5. Ortiz de Montellano, B. R. *Act. Congr. Int. Am. 42nd* **1976,** *6,* 287.
6. Ortiz de Montellano, B. R. *Estud. Cultural Nahuatl* **1980,** *14,* 287.
7. Ortiz de Montellano, B. R. *Act. Congr. Int. Am. 42nd* **1976,** *6,* 287.
8. Ortiz de Montellano, B. R. *Ethnomed.* **1975,** *3,* 249.
9. Ortiz de Montellano, B. R. *Science* **1975,** *188,* 215.
10. Garibay, A. M. *Tlalocan* **1946,** *2,* 235.
11. Ibid.
12. Aguirre Beltran, G. "Medicina y Magia"; Instituto Nacional Indigenista: Mexico, 1963.
13. Krickeberg, W. "Las Antiguas Culturas Mexicanas"; Fondo de Cultura Economica: Mexico, 1961; p. 125.
14. Ricard, R. In "The Spiritual Conquest of Mexico"; Univ. of California Press: Berkeley, 1966; pp. 187–93.

15. Aguirre Beltran, G., ibid., pp. 202–3.
16. Aguirre Beltran, G., ibid., pp. 132–48.
17. Comas, J. *Am. Indigena* **1954,** *14,* 327.
18. Foster, G. M. In "Colloquia in Anthropology"; Wetherington, R. K., Ed., Southern Methodist Univ.: Dallas, 1978.
19. Kay, M. A. *Ethnohistory* **1977,** *24,* 251.
20. de Esteyneffer, J. In "Florilegio Medicinal"; Anzures, C., Ed.; Academia Nacional de Medicina: Mexico, 1978.
21. Madsen, W. *J. Am. Folklore* **1955,** *68,* 123.
22. Viesca Trevino, C., personal communication, 1981.
23. Garibay, A. M. *Tlalocan* **1946,** *2,* 235.
24. Ortiz de Montellano, B. R. *Ethnomed.* **1975,** *3,* 249.
25. Ortiz de Montellano, B. R. *Science* **1975,** *188,* 215.
26. Coe, M.; Whittaker, G. In "Aztec Sorcerers in Seventeenth Century Mexico: The Treatise on Superstitions by Hernando Ruiz de Alarcon"; Institute of Mesoamerican Studies: Albany, 1982; p. 381.
27. Dibble, C.; Anderson, A. J. O., ibid., Book 10, p. 145.
28. Ibid., Book 11, p. 179.
29. Ibid., pp. 139, 146, 148, 162.
30. Ibid., pp. 139, 146, 148, 162.
31. Lopez Austin, A. *Estud. Cultural Nahuatl* **1972,** *10,* 129.
32. Hernandez, F. "Historia General de las Cosas de la Nueva España"; National Univ. Mexico: Mexico, 1959; Vol. 2, p. 348.
33. Motolinia, T. de B. In "Memoriales"; O'Gorman, E., Ed.; National Univ. Mexico: Mexico, 1971; p. 364.
34. de las Casas, G. In "Apologetica Historia Sumaria"; O'Gorman, E., Ed., National Univ. Mexico: Mexico, 1967; Vol. 1, p. 311.
35. de Esteyneffer, J., ibid., Vol. II, pp. 630, 635.
36. Patrick, L. L., Ph.D. Thesis, Univ. of Pittsburgh, 1977, p. 93.
37. Merkow, R., Ed. "The Merck Manual," 13th ed.; Merck: Rahway, 1977; p. 715.
38. Sociedad Farmaceutica Mexicana "Farmacopea Mexicana," 6th ed.; Editorial Botas: Mexico, 1966; p. 250.
39. Instituto Medico Nacional "Datos para la Materia Medica Mexicana"; Secretaria de Fomento: Mexico, 1894; Vol. 1, p. 368.
40. Perez-Cirera, R. *Gac. Med. Mex.* **1944,** *74,* 140.
41. American Herbal Pharmacology Delegation "Herbal Pharmacology in the People's Republic of China"; NAS: Washington, D.C., 1975; p. 117.
42. Gonzalves de Lima, O. "El Maguey y el Pulque en los Codices Mexicanos"; Fondo de Cultura Economica: Mexico City, 1956.
43. Pennington, C. W. *Econ. Bot.* **1958,** *12,* 95.
44. Wall, M. E.; Warnock, B. H.; Willaman, T. T. *Econ. Bot.* **1962,** *16,* 266.
45. Cruse, R. R. *Econ. Bot.* **1973,** *27,* 210.
46. Tschesche, R.; Wulff, G. *Fortschr. Chem. Org. Naturst.* **1972,** *30,* 461.
47. Vichkanova, S. A.; Goryunova, L. V. *Nauchn. Issled. Inst. Lek. Rast.* **1971,** *14,* 204; *Chem. Abstr.* **1973,** *78,* 155107z.
48. Majno, G. "The Healing Hand: Man and Wound in the Ancient World"; Harvard Univ. Press: Cambridge, 1975; p. 117.
49. Ibid., p. 118.
50. Cope, Z. *Med. Hist.* **1958,** *2,* 163.
51. Herszage, L.; Montenegro, J. L.; Joseph, A. L. *Bol. Trab. Soc. Argentina Cir.* **1980,** *61,* 315.

52. Chirife, J.; Scarmato, G.; Herszage, L. *Lancet* **1982,** *i,* 560.
53. Chirife, J.; Herszage, L.; Joseph, A.; Kohn, E. S. *Antimicrob. Agents Chemother.* **1983,** *23,* 766.
54. Bose, B. *Lancet,* **1982,** *i,* 463.
55. Majno, G., ibid., p. 117.
56. Davidson, J. R.; Ortiz de Montellano, B. R. *Ethnopharmacol.* **1983,** *8,* 149.
57. Malcolm, S. A.; Sofowora, E. A. *Lloydia* **1969,** *32,* 512.
58. Davidson, J. R., unpublished data.
59. "Compendium of Pediatric Skin Diseases"; Dome Chemical: New York, 1963, p. 9.
60. Fitzpatrick, T. B.; Eisen, A. Z.; Wolf, K.; Freedburg, I.; Austen, F. K., Eds. "Dermatology in General Medicine: Textbook and Atlas," 3d ed.; McGraw-Hill: 1979.
61. World Health Organization, *Medicina Tradicional (Mexico)* **1977,** *1,* 71.
62. Etkins, N. L. *Medical Anthropology* **1979,** *3,* 393.

0939-1/86/0022$06.50/0

Chapter 2

Zuni Indian Medicine: Folklore or Pharmacology, Science or Sorcery?

SCOTT CAMAZINE*

This chapter has its origins in a study of the traditional health care practices of the Zuni, a tribe of approximately 6500 American Indians living in a small community in west-central New Mexico. Field work conducted during the summers of 1977 and 1978 involved interviews with 27 medicine men and other Zunis, mostly older individuals *(1, 2)*. During the study 138 plant species were collected. For 49, the Zuni described a medicinal use. These plant remedies were examined in an effort to determine their pharmacological and physiological mode of action as well as their cultural significance.

THE ZUNI: HISTORY AND CULTURE

The ancestors of the Zuni are believed to have entered the southwestern United States as early as 10,000 B.C. Initially hunters and gatherers of wild plant foods, they developed agriculture in approximately 1000 B.C., growing squash, beans, and maize introduced from the southern Mexican plateau. The hunter-

*Division of Biological Sciences
Cornell University
Ithaca, NY 14853

gatherer existence was gradually replaced by small agricultural settlements that were consolidated during the Great Pueblo period between 1100 and 1200 A.D. The year 1538 marks the beginning of the Spanish influence, and the ensuing 200 years were a period of conflict during which the large pueblos were abandoned and present-day Zuni was established *(3)*. Aside from the conflicts and destruction, the most important consequence of the Spanish influence was the introduction of sheep and cattle, which further altered traditional subsistence patterns. The Zunis remained dependent upon the land for their livelihood until this century *(4, 5)*, but today most Zunis earn their living through the sale of turquoise and silver jewelry crafted at home. The land and its flora play a less important role in their lives, and nontraditional medicine provided by the U.S. Public Health Service has become widely accepted. As a result, knowledge of traditional herbal medicine has begun to wane.

TRADITIONAL ZUNI MEDICAL PRACTICES AND BELIEFS

Traditional medicine is practiced by the medicine man (called an *ak'wa:mossi* in Zuni), who is a member of one of the medicine societies. In its original form there were 12 such fraternities, each practicing general medicine but having specialties as well, such as cures for rattlesnake bites, sore throats, epilepsy, or bullet wounds. Folk remedies, largely herbal, are another form of traditional medicine, well-known to the elder Zunis and administered without seeking the aid of the medicine man.

Among the Zuni, natural causes, sorcery, disease-object intrusion, and breach of taboo are all regarded as origins of illness *(6–9)*. For common bruises, insect bites, or fractures, a supernatural cause is not sought and treatment is often in the form of herbal folk remedies. The treatment of an abscess with pinyon pine sap is an example.

> [But for] any complicated or mysterious trouble or one which does not yield readily to legitimate medicine, some higher power than man must be called upon to eradicate the disease 'shot' into the person by witchcraft. In such cases the Beast Gods act through their agents, the theurgists, who

> have great influence, the patient and the family showing every confidence in their doctor... *[6]*.

Today, the theurgist (or *ak'wa:mossi*) performs his rituals similarly to the way they were practiced prior to the advent of modern medicine at Zuni. For example, in the case of a gunshot wound with the development of a wound abscess, a medicine man visited the hospital. He sat beside his patient, who was lying in bed with his chest bared and the bandaged abdominal wound visible. The *ak'wa:mossi* opened a corn husk packet filled with white cornmeal and placed it on the floor. As he chanted in Zuni and moved his hands across the patient's chest, he materialized six small black pebbles one by one and placed them on the cornmeal. He later explained to me that these had been put there by someone who wished to harm the patient. These objects were preventing the recovery of his patient. Before the medicine man left, he gave the patient a bit of aromatic root, which they both chewed. This ritual appeared to alleviate the patient's anxiety. According to Bunzel *(9)*, the Zuni know that these "cures" are accomplished by sleight of hand. This knowledge in no way detracts from their effectiveness because the practices are sanctioned by the gods and the act itself symbolizes this relationship with the divinities.

The Zuni also believe in breach of taboo as a disease etiology. The occurrence of illness in these cases is related to beliefs in contagious magic and to the idea that "like affects like", as illustrated in the following examples. With contagious magic, some portion of the victim's body or clothing will be magically manipulated, and these actions result in misfortune or disease.

> Hair cuttings are burned. Were they thrown out, the winds would scatter the hairs and with them the life and fortune [of the child]...In washing a baby's clothes, much care is taken not to drop any garment—the child would have a bad fall *[10]*.

> A pregnant woman should not scatter bran on her oven floor—a method of testing temperature—otherwise her child will have pimples. Albinism is caused by a parent-to-be eating the white leaf inside the corn husk...During pregnancy a husband must be very circumspect in his treatment of animals. Were he to shoot a rabbit or prairie dog, the child

> would be marked or deformed in a way corresponding to the injury suffered by the animal—blind or lame or malformed *[8]*.

To what extent these beliefs are the concern of the Zunis today is difficult to judge. They are less prevalent among the more highly educated younger generation, yet among the older Zunis some of these beliefs persist.

THE ORIGIN OF ZUNI FOLK REMEDIES

How do folk remedies originate? Is there any scientific basis to the manner in which they are introduced? Zuni religion and folklore point to the gods as the source of their plant remedies. None of the Zunis interviewed could relate how any remedy came into being. One must assume that the process by which a specific plant comes to be used for a particular ailment is largely empirical; through the course of time, one can envision that people have tried nearly every conceivable means to affect the course of illness. Most likely, the search for plant remedies was not systematic. Furthermore, the evaluation of the results of a treatment probably would not meet today's medical criteria.

Folk remedies are introduced and persist in a culture in a multitude of ways. Many illnesses (e.g., viral syndromes) are self-limited, so that any remedy may be deemed "effective" when the illness subsides on its own over the course of several days. Therefore a great number of folk remedies are developed for self-limited diseases. Concomitants can be confused with causality in other ways. For example, Wallace *(11)* estimated that among 20% of the victims of poisonous snakebites, there is no envenomation. Predictably, therefore, the Zuni have many plant remedies for snakebites that they claim are effective. The Zuni have mistakenly attributed the lack of an illness to the "curative" effect of the remedy.

A certain logic to the origin of many folk remedies exists, but this logic would not pass the scrutiny of today's science. According to the doctrine of signatures, widely believed in many cultures, features in the appearance of a plant indicate its utility. Among the Zuni the milky sap of the spurge *(Euphorbia albomarginata)* is believed to promote the flow of milk in nursing mothers.

Similarly, the sticky leaves of evening-star *(Mentzelia pumila)* are thought to strengthen the grip and hold one firmly to one's horse while riding.

The primitive Zuni had no knowledge of the origins of disease. In an effort to attribute some connection between an ailment and its cure, the Zuni believed that the cure must be specific for and somehow related to the disease. To the unsophisticated mind, the doctrine of signatures, like affects like, would seem obvious. Other herbal remedies may have their origins from peculiar or unique properties of a plant. The highly aromatic plants in Zuni have medicinal uses. Did these uses develop because the aromatic essential oils have medicinal value or simply because these plants have unusual fragrances and as such are thought to possess special qualities?

Other plant remedies may have originated through an extension of the plant's initial use. The soft aromatic leaves of sagebrush *(Artemisia tridentata)* placed in shoes of sheepherders are well suited as a foot deodorant. In addition, the Zuni maintain that sagebrush prevents athlete's foot infection, an extension of its original use as a deodorant.

A second example is the use of sagebrush tea. The tea with its sharp camphor odor is drunk for symptomatic relief of nasal congestion associated with viral illnesses. By an extension of this use, many now believe that boiling the plant over the stove will prevent colds.

Certain plant remedies have gained favor because of their unpleasant taste or odor. Such remedies are as popular today as in the past. A medicine that is foul smelling or otherwise noxious is deemed powerful. The basis of this belief can be attributed to primitive man's conception of disease. Unable to discover any reasons for the cause of disease, primitive man attributed its devastating effects to malign influences, devils, demons, and the sorcery of enemies. An evil force was believed to enter the body of the ailing individual. The power of noxious remedies lies in their supposed ability to exorcise the malign influence. Ancient man thus utilized urine, excrement, onions, and smoke to create an unpleasant environment that would drive out the evil.

To summarize, folk remedies enter the cultural traditions of a people in various ways. Many folk remedies are certainly based upon primitive beliefs of the cause of disease, principles such as the doctrine of signatures, and a variety of folklore and superstitions. Nonetheless, serendipity and millennia of trial and error also result in the use of folk remedies with a valid scientific basis. For ethnobotanists, pharmacologists, and chemists, an initial task in the study of folk medicine is to unravel science from sorcery and folklore from pharmacology.

The discovery of herbal remedies such as digitalis, quinine, and ephedrine has provided a great impetus for the continuing isolation and characterization of plant secondary compounds. However, with well over 275,000 plant species worldwide, a random screening of plants for compounds with biologic activity is doomed to be frustrating and inefficient. Many workers have therefore chosen the ethnobotanic approach as a means of homing in on plants with pharmacologically active compounds of potential therapeutic interest.

THE SCIENTIFIC BASIS OF ZUNI MEDICAL PRACTICES: IS THERE ONE?

We can now ask, "What are the modes of action of traditional Zuni medicines? Is there any scientific basis to the many folk remedies of the Zuni?"

In this analysis, remedies are categorized into those with a physiological effect and those that appear to have only a psychosocial basis. In the physiological category are remedies that contain plant-derived products with pharmacologically active compounds. Other remedies, though containing plants, may not have their basis in herbal pharmacology, yet aspects of the remedy may have a physiological effect on the disease process. Finally, remedies may not appear to have any physiological effect. The rituals and ceremonies associated with folk medicine can be considered forms of counseling and psychotherapy. Remedies such as the removal of disease objects from the patient meet the social and cultural expectations of the patient and his family and may have a profound effect on the patient's sense of well-being. We cannot assume that such "remedies" have no physiological effects upon the patient.

Ample evidence shows that emotions dramatically influence physiological processes. Recent work has demonstrated an effect of stress upon the functioning of the immune system *(12)*, and a

National Academy of Sciences panel recently concluded that grief over the death of a family member may increase the risk, especially among men, of contracting an infectious disease or dying of a heart attack or stroke. One study has demonstrated that in hospitals a window view of a natural setting may have a beneficial effect on patients' recovery from surgery *(13)*. In this respect one must conclude that the folk medicines of the Zuni and all other cultures have a scientific basis. Remedies that meet the patient's expectations and relieve anxiety are likely to have both psychological and physiological benefits.

In the following sections, a number of case studies of Zuni folk medicine are presented and an attempt to evaluate the possible scientific basis of each therapy within the framework mentioned is made.

CASE STUDIES OF ZUNI FOLK MEDICINE

Abscesses

The resinous gum of the pinyon pine *(Pinus edulis)* is employed by the Zuni for the treatment of abscesses. In a typical case, the abscess is opened and sprinkled with the dry powdered resin, or the sticky gum may be mixed with lard and placed in the open wound, which is then bandaged.

The resin from the pinyon pine contains a volatile oil consisting of α-pinene (70–75%), β-pinene (5%), and α-cadinene (15–20%) *(14)*. The volatile oils have traditionally been used by many cultures for their antiseptic, rubefacient (counterirritant), expectorant, and diuretic actions. In a study of common antiseptics, turpentine (the volatile oil from species of pine, consisting largely of α-pinene and β-pinene) was shown to be effective in killing *Staphylococcus aureus,* a major skin pathogen *(15)*. The volatile oils are supposedly beneficial for chronic inflammations through their actions as weak antiseptics, mild analgesics, and chemotactic attractants for leukocytes *(16)*. Majno *(17)* examined the ancient medical practices of the Greeks, Romans, and Egyptians. He concluded that the use of tree resins, such as myrrh and frankincense, had a valid scientific basis attributable to their antibiotic effect.

Perhaps more important than any pharmacologic effect of the resins is the Zuni practice of opening the abscess. Incision and drainage of abscesses are standard medical practice today, and this procedure is usually all that is required to effect a complete cure.

Venereal Disease

Doveweed *(Croton texensis)* and thistle *(Cirsium ochrocentrum)* have been used to treat gonorrhea and syphilis. The remedy involves drinking a tea prepared from one of these plants and then running rapidly for about 1 mile to induce sweating. The patient must then bundle himself in blankets to further increase the sweating.

The physiological basis of this remedy may be the elevation of body temperature induced by vigorous exercise. When penicillin was introduced in 1943, it became the standard therapy for these venereal diseases, but as late as 1950 fever therapy was still being practiced *(18)*. Body temperature was increased to as much as 106 °F by means of injections of typhoid vaccine or malaria and by mechanical devices such as infrared head boxes, electric blankets, or hot baths of 112–15 °F. Exertion followed by the prevention of heat loss and cooling, as in the Zuni therapy, would be another effective means of raising body temperature. Running 3 miles on a hot day in less than 15 min is sufficient to produce a rectal temperature of 106 °F *(19)*. In vitro studies have demonstrated that the syphilis organism, *Treponema pallidum,* is immobilized by a temperature of 105.6 °F sustained for 2 h. In another study, 104 °F for 2 h killed the syphilis spirochetes in the external lesions of primary and secondary syphilis in man *(20)*. Substantial evidence indicates that the gonorrhea organisms are similarly sensitive to elevated temperatures *(21)*.

The use of an herbal tea prior to strenuous exercise is unlikely to have any pharmacologic effect but may serve to prevent dehydration. The chancres and skin lesions of syphilis subside spontaneously over the course of 2–6 weeks, but the organism still persists in the body to cause the late sequelae of the disease. Though elevated body temperatures may have a deleterious effect upon the infection, that this temperature elevation would completely eradicate the disease is doubtful.

Promoting Lactation

The leaves and roots of various species of spurge *(Euphorbia* spp.) are ingested to promote lactation in postpartum mothers. A great variety of chemical constituents have been isolated from various *Euphorbia* species, including flavonoids, amino acids, alkanes, triterpenoids, and alkaloids. The diterpenes in the latex are toxic, causing skin irritation and emesis and diarrhea when ingested. The following diterpenes have been isolated from *Euphorbia* species: phorbol esters, 12-deoxyphorbol, 12-deoxy-16-ingenol, 5-deoxyingenol, 20-deoxyingenol, resiniferotoxin, and tinyatoxin *(22)*.

That any of these compounds would promote lactation is doubtful. More likely the rationale and origins of this remedy are based upon the Zuni's belief in the doctrine of signatures. In this case, the milky latex of the spurges suggests their utility in promoting lactation. The nearby Ramah Navajos employ *Euphorbia serpyllifolia* for the same purpose *(23)*. The plant is ground and rubbed on the breast. Perhaps the topical application of this remedy has a physiologic basis because tactile stimulation of the nipple by suckling results in the ejection of milk through a hormonally mediated mechanism *(24)*.

Swellings

Five plants have been described as remedies for swellings of the body. Two species are in the genus *Oenothera (O. coronopifolia* and *O. hookeri)*. These members of the evening primrose family are used in a similar manner. Each evening saliva or water is rubbed on the swollen area and a powder prepared from the dried flowers is sprinkled on the moistened area so that the powder adheres. The blossoms of these flowers last a single day. In the evening they are large and succulent but by the next morning they wither. In a similar manner, the swellings of the body are supposed to disappear after using this medicine.

This description by a Zuni medicine man is based upon the doctrine of signatures. Likewise, the coyote-melon *(Curcubita foetidissima)* with its swollen fruits or deervetch *(Lotus wrightii)* with its pea pods may have also suggested their uses in the treatment of swellings.

Diabetes

Diabetes is a prevalent disease among the Zuni, affecting at least 25% of the population over the age of 45 *(25)*. It is almost exclusively of the adult-onset type. Twenty-two Zuni diabetics were interviewed to determine their beliefs about the illness and its therapy *(1)*. Three of the patients had been to the medicine man for the treatment of their disease. They were older individuals (67, 72, and 77 years old). Each had disease objects removed from his body in a traditional ceremony by the *ak'wa:mossi.* In one case an herbal tea was also prescribed. Eight of the 22 patients knew of a plant called kowahkyatsi that was supposed to be useful for diabetes, and one of these patients used the plant when it could be obtained. The plant, a thistle *(Cirsium ochrocentrum)*, has no known hypoglycemic activity. Its long tap root is dug from the ground and used fresh or stored for winter use. The root is boiled in a pot of water, and the resulting tea is consumed three times a day to "keep the sugar down". Two patients used another plant, a species of *Glycyrrhizia,* which was similarly prepared. The root was purchased in a shop in nearby Gallup, N.Mex. Its prominent constituent, glycyrrhizin, is extremely sweet. Once again, the use of this plant is probably based upon the doctrine of signatures, the sweetness of the plant suggesting its efficacy in treating a disease that results in excess sugar in the blood.

Rattlesnake Bites

Six plants were described in remedies for rattlesnake bites: *Helianthus annuus, Solanun elaeagnifolium, Grindelia aphanactis, Gaura parviflora, Astragalus amphioxys,* and *Croton texensis.* Numerous compounds have been isolated from these species, but their efficacy for treating snakebites is questionable. However, the use of these remedies does have a rational basis. When someone is bitten, the root is chewed by the medicine man, who then sucks upon the wound to extract the venom. More root is chewed and applied to the swollen area. The root is used singly or in conjunction with others. Because so many plants are regarded as effective for this ailment, the pharmacologic value of the plants must be questioned. That so many unrelated plants containing a variety of compounds would all be effective treatments for rattlesnake bites is unlikely.

The scientific basis of this remedy is twofold. First, 20% of all bites by poisonous snakes do not result in the envenomation of the victim. In the case of these "dry" bites, virtually any remedy would be effective. Second, the Zuni remedy calls for suction to the bitten area, which may remove some of the toxin. Topical application of a masticated root would be of doubtful value in counteracting the venom, but the tannins, resins, and other plant constituents may have some antiseptic or astringent properties.

Stomachaches

Fourteen plants were collected that are used for the treatment of stomachaches. Of interest from both a cultural and pharmacologic view are those plants that are said to induce vomiting. Although information is lacking with regard to some species in this category, most of these remedies are probably emetics.

A cultural expectation of the Zuni is that a treatment for stomachache induce vomiting. This expectation may have its basis in earlier times when certain gastrointestinal illnesses were probably caused by the ingestion of spoiled or toxic foods. In such instances, the prompt removal of the ingested substances may have prevented poisoning. Herbal remedies may have been used as we use syrup of ipecac today to induce vomiting for toxic plant and mushroom poisonings and drug overdosages. In addition, emesis often relieves, at least temporarily, the uncomfortable feelings of nausea that accompany certain gastrointestinal illnesses. A logical extension of this treatment would be to induce vomiting in an effort to cure the illness.

Of the plants used for stomachaches, many of the species (or closely related ones) are known to contain constituents that are emetic. The following is a description of the chemistry and pharmacology of some of these plants.

The leaves of *Croton texensis* are prepared as a tea for stomachaches. Plants in this genus, as many others in the Euphorbiaceae, are considered poisonous to livestock. *Croton tiglium,* the source of croton oil, was formerly used as a purgative but is now considered too drastic a medication. The oil is irritating to the gut, causing prolonged diarrhea and cramping as well as nausea and vomiting. Local application of the oil to the skin causes redness and blistering. Several drops of the oil are lethal to animals, causing a severe gastroenteritis *(26, 27)*.

Erysimum capitatum and *Dithyraea wislizeni* are members of the mustard family, Cruciferae. Mustard oil glucosinolates are present in many members of the mustard family and enzymatically yield pungent isothiocyanates when the plant tissues are chewed. A hydrolytic enzyme myrosinase is present in the plant but is sequestered from the glucosinolates, preventing hydrolysis until the plant cells are crushed in the presence of water *(28)*. The isothiocyanates are familiar as spices such as horseradish and mustard but in larger doses produce gastrointestinal irritation, nausea, and vomiting.

Eriogonum alatum, Erigonum jamesii, Atriplex argentea, and *Psoralea lanceolata* are members of genera that contain saponins. The saponins are named for their soaplike foaming quality which is a result of their characteristic structure. The saponin molecule is relatively large and has hydrophobic and hydrophilic portions, enabling it to form emulsions or colloidal suspensions. The saponins lower surface tension and lyse red blood cells when injected intravenously. Taken orally, they are poorly absorbed but are irritating to the gastrointestinal tract and cause vomiting.

Skin Lesions

More than a dozen plants have been used by the Zuni to treat skin lesions. In most cases the ailment is called a sore *(a'i:we)* or a rash *(su'do:we)*. On the basis of interviews with the Zuni, most of their remedies are used to treat minor abrasions, skin abscesses, and a variety of rashes. The abundance of remedies may be related to the self-limited course of many dermatologic conditions or to the variety of skin lesions that affect man.

Curled dock, *Rumex crispus,* is a species that has been used by many Indian tribes. The Paiutes and the Shoshones employ the plant for bruises, burns, swellings, and venereal disease *(29)*. The Ramah Navajos have used the plant to treat cutaneous and oral sores *(23)*. The Tarahumara of Chihuahua apply a poultice of the leaves and roots to sores on the feet and legs *(30)*. The Zuni use the ground-dried root to treat sores, rashes, and minor skin infections. Analysis of the root reveals emodin, chryosophanic acid, chrysarobin, tannin, volatile oils, resin, rumicin, sulfur, starch, and salts *(31, 32)*. Extracts of *Rumex crispus* have "on some occasions shown inhibition of *Staphylococcus aureus* and *Escherichia coli*" *(31)*.

Of most interest for their topical effects are chrysarobin and tannin. Chrysarobin (1,8-dihydroxy-3-methyl-9-anthrone) is fungicidal for *Trichophytum rubrum, Trichophytum mentagraphytes,* and *Microsporum canis,* organisms responsible for fungal infections of the hair, nails, and skin. Both chrysarobin and dithranol, a related compound lacking the 3-methyl group, have been used to treat psoriasis. The following description of the mode of action of dithranol is thought to apply to chrysarobin as well, and would explain their beneficial effect in the therapy of psoriasis, a skin disease in which the rate of epidermal growth is abnormally elevated *(33, 34)*.

Dithranol is a planar molecule and, like acridine derivatives and carcinogens such as benzpyrene, is thought to interact with DNA by intercalating between base pairs and interfering with protein synthesis. The DNA–dithranol complex may not be able to act as a template for de novo DNA synthesis or RNA synthesis, resulting in a cytostatic effect. The mitotic rate, determined by autoradiography, is decreased in the epidermis of guinea pigs following treatment with dithranol.

Tannins are topical agents that were widely used in a variety of situations earlier in this century *(35)*: (1) for treatment of inflammatory diarrheas, (2) for treatment of excessive secretion and swelling of the mucous membranes in oral illnesses, (3) for treatment of burns, (4) for prevention of bedsores, (5) as an antidote to various metal and alkaloid poisons, (6) as an astringent irrigation for the colon and vagina, and (7) as a suppository or ointment to shrink small hemorrhoids.

Taken internally as a colonic irrigation for the treatment of inflammatory diarrheas, tannins precipitate the proteins of mucus and surface epithelial cells, thus forming a mechanical barrier. The secretory activity and transudation of fluids in the gut are hindered, and the underlying mucosa may be protected from the effects of irritants and toxins in the bowel. Topical applications of tannic acid for burns rely on the same mechanism of action. Tannic acid was advocated for burns in 1925 and for almost 20 years was a favored treatment *(36)*. The objectives were "the prevention of absorption of toxins by converting injured tissue into inert tannates, and the formation of a firm eschar by precipitating protein to protect the tissue from bacterial infection and to prevent the extensive loss of fluid" *(37)*. Unfortunately, tannic acid was found to be hepatotoxic when absorbed in large quantities. Tannic acid enemas for diarrhea and the immersion of burn patients in tannic acid solutions resulted in the absorption of significant quantities of tannin through the inflamed gastrointestinal tract and denuded skin surfaces and were responsible for cases of severe centrolobular liver necrosis *(38)*. For this reason tannic acid therapy is now obsolete. Some species of *Lithospermum* and *Calliandra* contain tannins, which may account for the Zuni's use of *Lithospermum incisum* and *Calliandra humilis* as topical skin remedies. The dried root of *Rumex crispus* may contain as much as 6.4% tannin *(31)*.

A Zuni "Anesthetic" for Surgical Procedures

One of the most unusual and sacred Zuni plants is jimsonweed, *Datura inoxia*. The plant is used by the Zuni as a narcotic and hallucinatory medicine and may not be touched by anyone except a few medicine men and the rain priests. One medicine man refused to touch the plant, explaining that if he did, his skin would become covered with white splotches. In addition to its dramatic appearance, with large white tubular flowers, the plant contains the potent antimuscarinic alkaloid, atropine, which causes hallucinations and coma when taken in large doses.

Stevenson *(39)* describes its use:

> The writer observed the late Nai'uchi, the most renowned medicine man of his time among the Zuni, give this medicine before operating on a woman's breast. As soon as the patient became unconscious, he cut deep into the breast with an agate lance and, inserting his finger, removed all the pus; an anti-septic was then sprinkled over the wound, which was bandaged with a soiled cloth...When the woman regained consciousness she declared that she had had a peaceful sleep and beautiful dreams. There was no evidence of ill effect from the use of the drug.

> A small quantity of the powdered root...is administered by a rain priest to put one in condition to sleep and see ghosts.

> This procedure is for rain, and "rains will surely come the day following the taking of the medicine, unless the man to whom it is given has a bad heart."

Also, when a man is robbed he would summon to his aid a rain priest who would give him *Datura* so that he could see in his sleep the man who robbed him.

These effects of jimsonweed are almost certainly based upon the atropine content of the plant. A dose of 2 mg results in rapid heart rate, palpitations, dry mouth, dilated pupils, and blurring of the vision. Higher doses of 10 mg or more result in tachycardia, weak pulses, ataxia, restlessness, excitement, and, finally, hallucinations, delirium, and coma.

SUMMARY AND CONCLUSIONS

Studies of the plants used by the Zuni *(1, 2, 39)* document the use of at least 98 plants to treat nearly the entire range of human ailments. In addition to the important cultural and psychological value of these traditional Zuni medical practices, a number of the remedies also appear to have valid pharmacologic and physiologic effects.

Considering the great number of remedies and the likelihood that only a small fraction have a valid scientific basis, one naturally seeks means to increase the chances of finding potentially useful treatments. The following approaches may be useful.

First, an understanding of the culture and traditional medical beliefs may allow one to more readily cull those remedies that merely fulfill cultural expectations and have a psychotherapeutic basis.

Second, an understanding of the people also will prevent some of the ambiguity concerning how a particular remedy is used and for which ailment. For example, the Zuni word for rash, *su'do:we,* was applied to the gamut of conditions ranging from bacterial and fungal skin infections to psoriasis. A greater understanding of Zuni language and culture may have revealed more detailed and accurate diagnoses, which would have aided in the evaluation of these remedies.

Third, examination of only certain types of remedies, such as well-defined treatments that are used for specific ailments, may be most fruitful. For example, the Zuni know that jimsonweed is a powerful medicine that can be used to anesthetize patients and induce visions. In contrast, those plants that are used for diverse ailments and those ailments that are treated with diverse remedies fall into a different category, which may be less likely to yield useful remedies. Parallels with modern medicine exist. For conditions such as cancer, disagreement as to the best method of treatment occurs. Various surgical and medical approaches for many cancers exist because no single treatment is best nor very good. Similarly, many over-the-counter remedies are used for a variety of ailments because few have a substantial effect on the course of the disease; most are merely symptomatic treatments with questionable efficacy.

The study of folk medicine reveals much about a people, their beliefs, and their culture, and such studies have been pursued with great interest by anthropologists, botanists, physicians, and others. Whether continued studies in this area will lead to the discovery of valuable drugs or medical therapies remains to be determined.

LITERATURE CITED

1. Camazine, S. *Soc. Sci. Med.* **1980,** *14B,* 73–80.
2. Camazine, S.; Bye, R. A. *J. Ethnopharmacol.* **1980,** *2,* 365–88.
3. Foster, R. W. "Southern Zuni Mountains: Zuni-Cibola Trail"; State Bureau of Mines and Mineral Resources: Socorro, N. Mex., 1971.
4. Spencer, R. F. "The Native Americans"; Harper and Row: New York, 1965.
5. Workman, P. L. *Am. J. Phys. Anthropol.* **1974,** *41,* 119–32.
6. Stevenson, M. C. "23rd Annual Report of the Bureau of American Ethnology"; GPO: Washington, 1904.
7. Parsons, E. C. *Science* **1916,** *44,* 469.
8. Parsons, E. C. *Proc. 19th Int. Congr. Americanists* **1917,** p. 379.
9. Bunzel, R. "47th Annual Report of the Bureau of American Ethnology"; GPO: Washington, 1932; pp. 467–544.
10. Parsons, E. C. *New Mexico Man* **1919,** *19,* 168.
11. Wallace, J. F. In "Harrison's Principles of Internal Medicine"; Thorn, G. W.; Adams, R. D.; Braunwald, E.; Isselbacher, K. J.; Petersdorf, R. G., Eds.; McGraw-Hill: New York, 1977.
12. Shavit, Y.; Lewis, J. W.; Terman, G. W.; Gale, R. P.; Liebeskind, J. C. *Science* **1984,** *223,* 188–90.
13. Ulrich, R. S. *Science* **1984,** *224,* 420–21.
14. Guenther, E.; Althausen, D. "The Essential Oils"; Van Nostrand-Reinhold: New York, 1949.
15. Rose, E. S. *Am. J. Pharm.* **1929,** *101,* 52–55.
16. Sollman, T. "A Manual of Pharmacology," 3d ed.; Saunders: Philadelphia, 1926.
17. Majno, G. "The Healing Hand: Man and Wound in the Ancient World"; Harvard Univ. Press: Cambridge, Mass., 1975.

18. Weiss, R. S.; Joseph, H. L. "Syphilis"; Thomas Nelson: New York, 1951.
19. Newburgh, L. H. "Physiology of Heat Regulation and the Science of Clothing"; Hafner: New York, 1968.
20. Stokes, J.H. "Modern Clinical Syphilology," 3d ed.; Saunders: Philadelphia, 1944.
21. Carpenter, C. M. *J. Lab. Clin. Med.* **1933,** *18,* 981-90.
22. Kinghorn, A. D.; Evans, F. J. *Planta Med.* **1975,** *28,* 325-35.
23. Vestal, P. A. "Ethnobotany of the Ramah Navajo," papers of the Peabody Museum of American Archeology and Ethnology; Harvard Univ.: Cambridge, Mass., 1952; Vol. 40(4).
24. Goodman, H. M. In "Medical Physiology"; Mountcastle, V. G., Ed.; Mosby: St. Louis, 1974.
25. Long, T. P. *Am. J. Public Health* **1978,** *68,* 901.
26. Kingsbury, J. M. "Poisonous Plants of the United States and Canada"; Prentice-Hall: Englewood Cliffs, N.J., 1964.
27. Muenscher, W. C. "Poisonous Plants of the United States"; Collier Books: New York, 1975.
28. Van Etten, C. H.; Tookey, H. L. In "Herbivores: Their Interaction with Secondary Plant Metabolites"; Rosenthal, G. A.; Janzen, D. H., Eds.; Academic: New York, 1979.
29. Train, P. "Medicinal Uses of Plants by Indian Tribes of Nevada," Contributions Toward a Flora of Nevada #45, Nevada, 1957.
30. Bye, R. A., Jr., Ph.D. Thesis, Harvard Univ., Cambridge, Mass., 1976.
31. Watt, J. M.; Breyer-Brandwijk, M. G. "The Medicinal and Poisonous Plants of Southern and Eastern Africa," 2d ed.; Livingstone: Edinburgh, England, 1962.
32. Gibbs, R. D. "Chemotaxonomy of Flowering Plants"; McGill-Queen's University Press: Montreal, 1974.
33. Krebs, A.; Schaltegger, H. *Experientia* **1965,** *21,* 128-29.
34. Swanbeck, G.; Liden, S. *Acta Derm. Venereol.* **1966,** *46,* 228-30.
35. Blumgarten, A. S. "Textbook of Materia Medica, Pharmacology, and Therapeutics," 7th ed.; Macmillan: New York, 1943.
36. Davidson, E. C. *Surg. Gynecol. Obstet.* **1925,** *41,* 202-21.
37. Goodman, L. S.; Gilman, A. "The Pharmacologic Basis of Therapeutics," 2d ed.; Macmillan: New York, 1960.
38. Goodman, L. S.; Gilman, A. "The Pharmacologic Basis of Therapeutics," 5th ed.; Macmillan: New York, 1975.
39. Stevenson, M. C. "30th Annual Report of the Bureau of American Ethnology"; GPO: Washington, 1915, 1908-9.

0939-1/86/0039$06.00/0

Boerhaavia diffusa (Reproduced with permission. Copyright 1978 Reference Publications, Inc.)

Chapter 3

Ayurveda: The Traditional Medicine of India

NARAYAN G. PATEL*

According to an ancient Indian legend, a learned guru named Atreya had a student, Jivaka. After Jivaka had studied medicine with Atreya for many years, he asked his teacher if he could go to the people and help them. Atreya said, "Let me give you a final test. Go out and search an 8-mile area around our ashram and collect all plants that have no medicinal value." The student went out and examined every plant he found. He tasted the fruits of some, and he examined the leaves and flowers of others. Finally he came back, bowed down, and sorrowfully said, "I could not find a single plant that does not have medicinal value." Atreya was very pleased, blessed Jivaka, and told him that he was now ready to go to the people and help them. This lesson of 3000–4000 years ago has value for us today.

Knowledge of the medicinal value of plants is recognized by almost every society on earth. Once a specific plant was discovered to be therapeutic, this knowledge was passed on to others as folk medicine. In India, a society with a strong cultural heritage, this knowledge attained a well-organized form. The knowledge was systematically recorded and employed as a traditional health care system called *ayurveda*. Ayurveda literally means knowledge

*Agricultural Chemicals Department
Experimental Station
E.I. du Pont de Nemours & Company, Inc.
Wilmington, DE 19898

of life. Ayurveda collectively encompasses medicinal, psychological, cultural, religious, and philosophical concepts. It is a holistic approach that has as its goal a long, healthy, and happy life. Today, some 70–80% of the people in India follow the ayurvedic approach to health care. Other systems include unani-tibb (Greco-Arabic), siddha (southern Indian, Tamil), allopathy (Western), naturopathy, and homeopathy.

The knowledge of the traditional medicine of the Indian subcontinent has been accumulated during 3000–5000 years. Early records of sporadic observations on the efficacy and the therapeutic values of the plants are found in *vedas,* the books of knowledge. More detailed compilations of observations were systematically done and recorded later. This knowledge is documented in *samhitas,* the compilations, which present the science of ayurveda. Two such samhitas, "Charak" *(1, 2)* and "Sushrut" *(3, 4)*, are the cornerstones of ayurveda. The "Charak Samhita", with commentaries, is in five volumes with 120 chapters covering more than 4000 pages. This treatise elaborates on eight major branches of human health, namely, (1) internal medicine *(kayachikitsa),* (2) surgery *(shalya tantra),* (3) psychology *(grahachikitsa/bhut vidya),* (4) toxicology *(agad tantra),* (5) oto-rhino-laryngo-ophthalmology *(urdwangchikitsa),* (6) pediatrics *(balachikitsa),* (7) eugenics and aphrodisiacs *(vajikaran tantra),* and (8) geriatrics *(rasayan tantra).* "Sushrut Samhita" is a masterpiece of eight chapters and 186 sections. It describes details of anatomy, discusses aspects of physiology, and delineates detailed surgical maneuvers with specific instruments.

According to ayurveda *(1, 3, 5–8)*, the human body is composed of five primary elements *(mahabhutas),* namely, earth (matter, *prithvi*), water *(ap),* light *(tejas),* air *(vayu),* and ether (space, *akasha*). The body's physical form is attained by a combination of these elements and is manifested by seven tissue identities: plasma *(rasa),* blood *(rakta),* muscles *(mamsa),* fat *(meda),* bones *(ashti),* red-yellow marrows *(majja),* and semen *(sukra).* Ayurveda views human beings as complex psychosomatic entities exhibiting a particular body constitution *(prakruti)* resulting from the three *doshas.* In the tri-dosha concept, the equilibrium of *vata, pitta,* and *kapha* prakruties is necessary for normal health. Any deviation or imbalance causes illness.

Vata depicts fluidity and motion in the body. Air and space constitute vata tendencies. Vata involves secretion, excretion, movement of air (as in breathing), and transport of neural energies via sensory and motor functions. Vata ailments include rheumatic arthritis, asthma, backaches and some headaches, constipation, sciatica, varicose veins, sexual dysfunctions, and menstrual problems. Imbalance of the vata dosha will cause worries and fear.

Pitta deals with energetics. Conceptually, fire and water constitute pitta tendencies. Metabolism, digestion, hunger, thirst, and thermoregulation fit in this category, and disorders like peptic ulcers, colitis, hyperthyroidism, hemorrhoids, and migraine headaches are considered pitta disorders. A pitta person is very active and is prone to anger and hate.

Kapha expresses a static and passive constitution. Earth and water constitute kapha tendencies. Kapha influences membrane functions, coherence of body organs and tissues, and homeostasis. Illnesses related to joints and their lubrication, accumulation of fat, bronchitis, emphysema, asthma, throat troubles, gastrointestinal ailments, and diabetes predominate in a kapha person. Psychologically a kapha person has tendencies toward possessiveness and greed.

A disease will manifest different symptoms in persons with vata, pitta, or kapha constitutions. Therefore, the treatment is varied. Also, the composition of the vehicle for a drug is chosen to suit the dosha for effective cure. For example, a vata person will be given medication with oil, pitta with ghee (boiled, clarified butter), and kapha with honey.

In understanding and appreciating ayurveda, one must remember that ayurveda is a traditional medical system. It takes the whole human body into account in assessing illness. Ayurveda rarely treats the symptoms but attempts to cure the disease on a permanent basis. This health care system acknowledges that some diseases may be caused by invisible minute life forms (germs). Ayurveda does not treat such patients, however, with harsh chemicals to destroy causal organisms. Instead, it tries to enhance immunity by medication and dietary approaches.

Many intriguing aspects of ayurvedic medicine deserve special mention. These aspects include, for example, formulations such as quaths, bhasmas, ghritas, and churnas, each meticulously prepared to increase the bioavailability of the active ingredients, protect them from degradation, and enhance their efficacies. Synergism is another aspect of ayurvedic preparations that is apparently achieved by selectively blending many plants, minerals, and animal products and thus maintaining the active ingredient at a minimum level and reducing or eliminating its side effects. In these preparations numerous molecules are present, each with a specific biological activity. These preparations could provide limitless opportunities to discover novel molecules and explore their mechanism of action. The dietary restrictions during medication raise questions of the compatibility and antagonism of the medication and the food. Above all, the belief component seems to play a crucial role in the recovery process. This phenomenon is not well studied. Significant contributions could be made by understanding the brain's neurochemicals and learning if the neurochemicals in any way participate differently in the recovery of a believing patient.

Vaidyas (ayurvedic doctors) diagnose an illness by ascertaining the body's balance (health) or imbalance (sickness) and which of the vata, pitta, or kapha doshas is affected and by how much. The feel of the pulse, *nadi,* is a crucial diagnostic tool. The art of feeling the pulse has attained extreme sophistication in ayurveda. The pulse's form, mode, intensity, and rhythmic frequency are felt externally by an experienced vaidya and are believed to reflect the performance of internal organs and maladies and imbalance in the three humors. The subsequent treatment proceeds to restore the balance of the whole body rather than suppress the symptoms. In ayurveda, as in other disciplines, the efficacy of the treatment for the desired cure is of paramount importance. Many diseases are self-limiting and get cured by themselves. Therefore, evaluating how much a physician and medication really help is a complex task.

Three specific conditions, the common cold, diabetes, and leukoderma, have been selected for discussion.

THE COMMON COLD

For a patient with a common cold, the vaidya employs experience and his subjective evaluation. He considers the prakruti of the patient to catch a cold. The vaidya feels the pulse and asks a few questions to determine what the patient did to precipitate the cold. The vaidya prescribes herbal tea, lots of fluids, rest, meditation, alterations in daily routines, and abstention from certain foods. The patient believes the vaidya will cure him. The patient gets well at little or no cost.

For a similar patient, an allopathic doctor takes the pulse, checks the temperature and respiration, uses the stethoscope, establishes the severity, recommends rest and lots of fluids, and prescribes analgesic and antipyretic pills. The patient believes the medication will cure him. The patient gets well at a moderate charge.

This comparison illustrates the similarities and differences in diagnosis and treatment of one of the many self-limiting ailments. The physician's help may have prevented the patient from becoming seriously ill, but the body's own defense mechanisms were of greater importance in restoring health.

A CASE OF DIABETES MELLITUS

Diabetes [*madhumeha,* meaning honey urine *(9, 10)*] is a complex metabolic disorder. Historically, each major new finding has cast some doubts on the earlier understanding of the disease and simultaneously given new insights that subsequently have been challenged. The early model of diabetes as dysfunctions of metabolism and transport of glucose is no longer adequate. We now know lipids, proteins, amino acids, ketones, and hormones are involved. The pancreas was recognized early as the primary organ involved. Now the liver, muscles, adipose tissues, and the brain are part of the total picture *(10–13)*.

Allopathic treatments are aimed at reducing the level of blood sugar. These treatments are done mainly by injection of insulin and in some instances by oral hypoglycemic agents *(13, 14)*. No attempt is made to restore the pancreas to normal functioning. Ayurvedic treatments are all oral. Hence, if the medicine contains an active ingredient (AI), it must survive digestive degradation. Another possibility is that an inactive form may be converted to the AI. These treatments are intended to restore permanent cures rather than provide metabolic relief *(9, 15)*.

As an example of how ayurveda treats diabetes, a case is considered. A 25-year-old male patient arrives. The patient is exhausted. The vaidya knows him and his family. The family has madhumeha tendencies. The vaidya is aware of the dietary influence on this condition. The vaidya takes the case history, asking questions about exhaustion, excessive thirst, the volume of urine, skin eruptions, the amount of exercise, and the patient's daily rituals. The vaidya tells the patient that the symptoms indicate mild madhumeha. According to "Madhavnidan" *(9)* (an ayurvedic disease-diagnostic text), the disease is difficult to cure, but nonetheless relief is possible. Ayurveda classifies diseases as *susadhya,* easy to cure, *kasadhya,* difficult to cure, or *asadhya,* incurable.

The patient mentions that his friends suggested he eat two tender leaves of neem *(Azadirachta indica)* and bilva *(Aegle marmelos)* on an empty stomach every day *(15)*. This treatment *(16)* works, says the vaidya. He knows that scientists have investigated both plants for their chemical constituents. However, there are no clinical studies on these plants' hypoglycemic properties.

The vaidya can choose to prescribe *dhanvantari ghrita, trivanga bhasma,* or *arogyavardhini gutika (17)* for a possible long-lasting cure. Whatever he prescribes, he warns the patient that for good results strict dietary requirements are crucial. The patient must avoid all starchy foods, sugar in any form, sweet fruits, butter, and ghee. Oil in small quantities is permissible as well as the sweet Jamun fruits *(18) (Euginia jambolana,* syn. *Syzygium cumini)* and *karela* (bitter ground, *Momordica charantia*); they reduce urine sugar *(15, 19)*.

Dhanvantari Ghrita

Ghrita is formulated with decoctions of plant juices dissolved or suspended in ghee. These preparations are primarily lipoidal, but due to ghee's solubility characteristics considerable amounts of polar constituents are also retained. These properties make ghrita easily assimilable when administered orally or even when applied to the skin. Dhanvantari ghrita is prepared according to *Astangahridaya chikitsasthan* (*Adhyaya* 12:9–22) and contains ingredients from 38 plants, water, and ghee *(17)*. Different plant parts, e.g, roots, rhizomes, heartwood, fruit pulp, seeds, and stems of specific plants are blended and processed in the required amounts.

In recent years these and other plants have been tested for hypoglycemic activity. The data were compiled in a review article by Chaudhary and Vohra *(20)*. The efficacies were ascertained either clinically in diabetic patients or with chemically *(alloxan)* induced diabetes in test animals. From these experiments the following plants were considered to have hypoglycemic activity: *M. charantia, Gymnea sylvester, Pterocarpus marsupium, Tinaspora cordifolia, Eugenia jambolana, Coccinia indica, Caseria esculenia,* and *Clerodendron phlomidis (16, 21)*. Chemicals isolated from these and other indigenous plants were tested. Cedrilone isolated from cedar wood demonstrated moderate hypoglycemic activity and low mouse toxicity (1500 mg/kg, orally). In addition, three nonplant preparations, shilajit, trivanga bhasma, and jasad bhasma, were found to be efficacious *(20)*.

Trivanga Bhasma

Bhasmas are nonplant ayurvedic preparations *(17, 22)*. Chiefly, bhasmas are composed of metals, but in a few instances they contain marine and animal products. Bhasmas are converted to a powder form by calcination. Calcination is achieved by blending the metal with specified plant juices and heating in a specially prepared earthen crucible. Calcination must be done with cow dung cakes in an earthen pit that avoids the excessive heat that is believed to destroy the medicinal properties. This calcination process is known as *puta.* A bhasma may require 10–1000 putas.

Before calcination, the metal must go through two main processes, physical purification *(sodhana)* and chemical detoxification *(marana),* to achieve the desired biological-medicinal properties. A bhasma is suitable for drug use if it has no metallic luster, is extremely fine such that the particles are smaller than skin ridges of the index finger and the thumb, floats on cold water, does not revert to metallic form, maintains its potency indefinitely, and manifests no toxicity. The greater the number of putas, the less the toxicity and the higher the efficacy.

The *trivanga bhasma* is prepared according to "Siddhayoga Sangraha" *(17)*, an ayurvedic text. The bhasma contains equal proportions of lead, tin, and zinc. Clinically, trivanga bhasma is efficacious and is presumed to induce insulin secretion *(20)*. Investigation of the site of action and mechanisms of secretion induced by trivanga bhasma would be interesting.

Many bhasmas are available for many ailments. Bhasmas are intriguing formulations of metals apparently associated with organic ligands. One is tempted to postulate the fate of the metal in ayurvedic bhasmas. Perhaps the bhasma binds to a carrier macromolecule, acts as a catalyst, or alters the membrane fluidity. The role of the plant juices in formulation also remains to be investigated. Apparently, the organic ligand renders the metal easily assimilable. A serious scientific investigation may uncover novel principles in metal therapy or warn against harmful uses. Several major bhasmas and their uses are listed in Table I.

Although honey is one of the most frequently suggested vehicles in ayurvedic texts, royal jelly, another honeybee product, is apparently not mentioned. Natural food users claim that royal jelly very quickly lowers blood sugars when taken orally by diabetic patients. In fact, it does contain a polypeptide that is similar to bovine insulin *(23)*. The crude royal jelly and a fraction that comigrates chromatographically metabolize [^{14}C]glucose in

Table I. Some Common Bhasmas, Their Main Ingredient, and Their Possible Uses

Sanskrit Name	*Ingredient*	*Uses*
abhrak	mica	anemia, anorexia, eye disorders
kasisa	iron sulfate	anemia, stomach disorders, spleen enlargement, hiccups, diabetes
tamra	copper	stomach disorders, ascites, inflammation, tissue disorders, eye disorders, leprosy
naga	lead	diarrhea, spleen enlargement, diabetes
parad	mercury	syphilis, genital disorders, rejuvenation
malla	arsenic	asthma, leukoderma, paralysis, impotency
mukta	pearl	cough, impotency, eye disorders, tuberculosis, sprue
yasada	zinc	dysentery, sweating, phthisis, tuberculosis, diabetes
rajat	silver	wasting, nerve disorders, brain functions, eye disorders, tuberculosis
loha	iron	sprue, stomach disorders, anemia, diabetes, blood disorders
vang	tin	asthma, cough, sweating, blood disorders, diabetes
svarna	gold	sprue, anemia, wasting, tuberculosis, muscle disorders, syphilis
hiraka	diamond	cancers

in vitro incubation with rat adipose fat tissues. The molecule is about 5000–6000 daltons, contains disulfide bonds, and has an amino acid composition very similar to that of bovine insulin (N. G. Patel, unpublished data).

A CASE OF LEUKODERMA (VITILIGO) *(24–27)*

A 29-year-old woman quietly enters a vaidya's reception room where a mother and a bright-eyed little girl are waiting. The little girl glances at the woman and whispers to her mother about her pretty sari. The woman, apparently upset and annoyed, turns her head away from them. She thinks that the girl is making fun of her ugly face. The woman mumbles, "Why don't people leave me alone? Why do they always talk about *kodha* (white skin patches) on my face?" She has one large patch over her forehead near the hairline and a couple on the side below the ear and spreading over the cheek. This cosmetically unpleasant, psychologically devastating skin depigmentation disease is known as leukoderma. The myth that the disease is contagious and incurable has made the woman and other leukoderma patients withdrawn and mentally depressed. The woman suffers from an extreme inferiority complex even though she is well educated.

The vaidya sees the woman next, in privacy, trying to gain her confidence. The vaidya asks a few simple and sympathetic questions without mentioning her depigmentation directly. The woman opens up and tells the vaidya that her life is unbearable. Family members and friends are unfriendly and avoid her as if she is untouchable. They are afraid that they will catch it. The woman tells the doctor that now her family does not allow her to eat dinner with them. The woman has to disappear if someone visits the family. The woman has to keep her dishes separate and also wash her clothes separately. "Today I am driven to the edge," she says. "I have already gone to Western–allopathic doctors, a homeopath, and one vaidya. It seems there is no recourse but to burn myself with kerosene if something is not done for my kodha. Why has God singled me out?" the woman sobs. She has been religious, has lived a virtuous life, and has never allowed any man to touch her. "Except," the woman confesses, "when I was 18, one man planted a kiss on this cheek, where the first patch started. Now I am an imperfect woman."

With deep sympathy the vaidya explains to her that the kiss had no connection with the disease. In fact, nobody knows for sure why one gets the disease, but the important point is that now there is hope that the disease can be cured. The vaidya tells the woman that his friend and a colleague *(25)* have studied psychological and sociological aspects of this dreaded disease in some 500 patients. It affects both sexes equally, though more females seek medical treatment. They also found that 232 patients were affected emotionally, and 39 women had marital problems. Out of 30 women who were cured, 15 got married and started a family. With this news, a little sparkle glitters in her eyes and her cheeks swell up and lips stretch with a smile.

This is a typical case of leukoderma, which is a social taboo, though not a health hazard.

The vaidya examines the woman's white patches closely and keeps asking questions about her eating habits. The vaidya asks the woman to bring in a stool sample that can be examined for a helmintic, protozoan, or an amoeboid infection. After the tests are completed, the vaidya gives the woman several pills of ayurvedic medication to take internally and medicated oil to apply externally, and then the vaidya tells the woman to stay in the sun every morning for 5–10 min. The woman is warned that success of the treatment will depend on how rigorously she follows the dietary restrictions. The woman should not eat fish, milk, curd, or radish or take allopathic antibiotics. She must follow the medication without interruption. The vaidya assures the woman that there is a good chance she will get cured. She should have a very positive attitude toward life and she should be thankful that she is very healthy and quite pretty and the disease is now curable *(26, 28–32)*.

Leukoderma is prevalent in the tropics and in India. Leukoderma's rate is very high with some 60,000 cases in the urban and surrounding area of Medabad, Gujarat State. People living on the farmland and in villages are not affected. No confirmed etiological factors or attributes for the cause of leukoderma can be found *(33, 34)*. Stomach parasites like helminths, protozoa, and amoebas are often found in leukoderma patients. Also, digestive disorders; constipation; irregular eating habits; simultaneous consumption of fish, milk, and eggs; and indiscriminate use of antibiotics are among other factors suspected to be involved in leukoderma incidence *(33, 35)*.

Most of the depigmentation is bilaterally distributed. The areas affected in order of increasing frequency are legs, feet, knees, lips, soles, back, chest, ankles, forearms, and face. In a study of 6700 patients, the repigmentation frequency was 100% in 81 cases, 71-99% in 622 cases, 50-70% in 1774 cases, and less than 50% in 2459 cases. In 1764 cases the patients discontinued the treatment *(36)*. The most responsive areas are, in decreasing order, forehead, back, face, eyelids, elbows, scalp, chest, legs, knees, and forearms. The sole regions do not respond at all, and little response is observed in the ankles, fingers, and palms. It is difficult to correlate the success rate of repigmentation in relation to the area of depigmentation.

Ayurveda "Charak Samhita" describes skin as having seven layers, which correlates well with modern histological observations (Table II). *Tamra* (coppertone) and *vedini* (sensitive to pain) layers are mentioned to contain dark granules of pigments. These layers fairly accurately describe where subcellular localization and release of melanocytes occur and where nerve endings terminate. The tamra color reported in ayurvedic texts could be correlated with the color changes during premelanosome maturation into darker melanosomes *(37)*. The leukoderma *(sweta kusta* or *swirta)* is literally called white leprosy *(28, 29)*. Ayurveda considers digestive disorders the cause. Hence, the first step in the curative regimen would be with kutaja or kado bark *(Holarrhena*

Table II. Comparison of Ayurvedic and Allopathic Observations of Mammalian Skin

Ayurvedic	*Allopathic*[a]
	Epidermis
abavasani (cover)	stratum corneum—horny, keratinized, oily
lohita (blood)	stratum lucidum—homogeneous, proteins, phospholipids
sweta (white)	stratum granulosum—granular, keratinized
tamra (copper)	stratum spinosum—prickle cell, SH, S-S, tonofibril
vedini (plain)	stratum germanatum—basal, melanocytes
	Dermis
rohini (red)	pars pappilae—cementing
mansadhara (muscular)	pars reticularis—connective tissues, inclusions

[a]Characteristics and properties.

antidysenterica) and mhallatka or bhilama *(Semecarpus anacardium),* which are vermifuges. Modern surveys also indicate stomach-related factors. Other ayurvedic treatments include oil of bakuchi *(Psoralea corylifolia),* khadira *(Acacia catechu),* and amlaki *[Emblica officinal (28)].* All treatments are to be followed by exposure to sunlight.

Of particular interest is the use of kakodumber *(Ficus hispida)* fruit powder, taken internally and mixed with gud (jaggery) and applied externally while the patient is exposed to sunlight for the desired efficacy *(38)*. Most of these plants, for example, *Ammi majus* [unani medicine *(30, 32)*], *P. corylifolia,* and *Ficus* spp., contain psoralen (N. G. Patel, unpublished observations), which, when photoactivated with UV, nicks DNA. The gud solution applied topically may effectively filter in long wavelengths to photoactivate psoralen. The phototoxic properties of *Ficus hispida* are also examined *(39, 40)*. Ayurveda recommends the use of tamra *(copper)* bhasma for leprosy *(kusta)* in addition to other plant preparations *(18, 22)*. The ayurvedic knowledge accumulated over centuries of numerous therapeutic properties of plants is unsurpassable *(41–49)*. Plants collected during the 1983 expedition and their selected uses are listed in Table III; and treatments of leukoderma are listed in Table IV.

The unani medicine has recognized *bars* (leukoderma) and has found successful treatment with aatrilal, *Ammi majus (30, 32)*. Aatrilal contains psoralen and is presumed to be involved in curative properties and in melanization *(24, 34, 35, 50)*, although the precise mechanism of its chemical reaction in the biochemical pathway is not understood.

Siddha medicine is unique in its uses and recommends copper-containing herbal preparations for leukoderma treatments *(51)*. "Copper plants" that retrieve copper from rocks and soils are used. One such plant, etti *(Strychnos nux-vomica),* contains as much as 0.24% copper. Copper binds to the cofactor essential for tyrosinase activity in melanin formation. The internal and external uses of copper plants are logical. Reducing agents such as thiol groups and ascorbic acid inhibit melanosis. Copper and other metals bind SH and reverse inhibitory action. Tamra (copper) bhasma reduces the concentration of ascorbic acid in adrenal glands in rats and guinea pigs *(52)*. The high serum copper levels are observed in

Table III. Ayurvedic Medicinal Plants: India

Plant	*Uses*
Abrus precatorius	purgative, emetic, toxic, aphrodisiac, treatment of nerve and skin disorders, abortifacient
Abutilin indicum	antigonorrheal, demulcent, astringent, diuretic, aphrodisiac, laxative, antihemorrhagic, antipyretic, treatment of bladder and urethra diseases
Acacia arabica	strengthens teeth and gums, promotes wound healing, prevents miscarriage, antidysenteric, galactopoietic, influences fetus skin pigmentation
Acacia leucophloea	astringent
Achyranthes aspera	diuretic, antihemorrhagic, treatment of boils and skin eruptions, antitussive, antiasthmatic, treatment of enlarged spleen (in malaria), analgesic
Acalypha indica	analgesic, antitussive, antiemetic
Adhatoda vasica	antitussive, antiasthmatic, antispasmodic, insecticide
Aegle marmelos	laxative, astringent, digestive aid, carminative, antidiarrheal, antidiabetic, antipyretic, antidynsenteric, treatment of sprue
Aerva lanata	anthelmintic, diuretic, demulcent, analgesic, kidney stone decalcification, treatment of jaundice
Ailanthus excelsa	antipyretic, expectorant, antiseptic, antiasthmatic, stimulates plant growth
Alangium lamarckii	lowers blood pressure, anthelmintic, purgative, treatment of leprosy and skin diseases, parasympathetic stimulation, emetic, stimulates plant growth
Alangium salvifolium	purgative, anthelmintic, treatment of skin diseases, treatment of rabies
Albizzia lebbeck	antiinflammatory, antihemorrhoidal, treatment of internal bleeding, treatment of snakebite and scorpion sting, astringent, antidiarrheal, treatment of night blindness and eye troubles, treatment of lymphoma, antiscrofulous
Amaranthus spinosos[a]	antimenorrhagic; antigonorrheal; antieczematous; lactational; emollient; poultice to abscesses, boils, and burns; kidney stone decalcification
Andrographis paniculata	antipyretic, treatment of liver troubles
Anisomelis ovata	treatment of sinus congestion, analgesic, antipyretic, antirheumatic
Argemone mexicana	treatment of skin diseases, laxative, expectorant, diuretic, treatment of jaundice, treatment of snakebite
Argyreia speciosa	antirheumatic, treatment of mental dullness, treatment of infirmity of old age, poultice for abscesses
Artemisia vulgaris	emmanogogue, anthelmintic, antispasmodic, treatment of stomach disorders, antiasthmatic, treatment of diseases of the brain

Continued on next page

Table III. *Continued*

Plant	*Uses*
Asparagus racemosus	demulcent, galactopoietic, antidysenteric, antidiabetic, diuretic, aphrodisiac, anticonvulsant, tranquilizer, treatment of colic
Azadirachta indica	astringent, treatment of antiperiodic boils and skin diseases, treatment of snakebite and scorpion sting, insecticide
Balanites aegyptica	purgative, expectorant, antitussive, treatment of colic, anthelmintic, as fish poison
Baliospermum montanum	purgative, stimulant, treatment of snakebite, cathartic, diuretic, treatment of jaundice, antiasthmatic
Barleria prionitis	antitubercular, antiseptic, antitussive, analgesic, treatment of boils and glandular swellings, prevents bleeding
Bauhinia racemosa	promotes urination, analgesic, antihelmintic, antihemorrhagic, antimalarial, fibers for sutures
Bauhinia vahlii	antitubercular, antiscrofulous
Bauhinia variegata	astringent, treatment of skin diseases, antiulcerative, antihemorrhagic, anthelmintic, treatment of snakebite, antidiarrheal, antiflatulance, antiscrofulous, treatment of blood and lymphoid disorders, antibacterial in ear inefections
Berberis aristata	treatment of urinary diseases; treatment of skin diseases; treatment of eye, ear, and mouth diseases; antimalarial; treatment of liver and spleen diseases
Boerhaavia diffusa	diuretic, expectorant, antiasthmatic, antiinflammatory, treatment of snakebite, treatment of hepatic disorders, antirheumatic, treatment of gout
Bombax ceiba	treatment of acne, aphrodisiac, treatment of menorrhagia, antidiarrheal, antihemorrhagic
Boswellia serrata	diaphoretic, diuretic, astringent, antirheumatic, treatment of nerve and skin diseases, treatment of laryngitis, antiseptic, antitussive
Bryonopsis laceniosa	induces maleness to embryo, treatment of leprosy
Bryophyllum calycinum	treatment of bruises, wounds, boils, insect bites; bone setting; kidney stone decalcification
Butea monosperma	treatment of leprosy, treatment of psoriosis, astringent, diuretic, anthelmintic
Caesalpinia pulcherima	antipyretic, antiasthmatic, antiinflammatory
Calotropis gigantia	anthelmintic, antiscabietic, antieczematous, treatment of elephantiasis, diaphoretic, expectorant, antipyretic, treatment of visceral enlargement, treatment of leprosy, antiasthmatic
Cardiospermum halicacabum	antirheumatic; treatment of stiffness of limbs, snakebite, and nerve diseases; analgesic
Cassia angustifolia	blood purifier, treatment of leprosy, treatment of habitual constipation

Table III. *Continued*

Plant	*Uses*
Cassia auriculata	treatment of skin diseases, anthelmintic, treatment of bruises, treatment of ophthalmitis and conjunctivitis, antidiabetic, treatment of chylous urine
Cassia chundra	treatment of skin diseases and leprosy
Cassia fistula	laxative, antirheumatic, treatment of snakebite, astringent, treatment of leprosy and skin diseases, treatment of gastric complaints, expels bedbugs
Cassia occidentials	antiasthmatic, antitussive, cures hiccups, antieczematous, treatment of scorpion sting, diuretic, treatment of elephantiasis, treatment of lung cancer
Cassia tora	treatment of skin diseases, anthelmintic, relief of itching, analgesic
Celastrus paniculta	treatment of mental disorders and schizophrenia; tranquilizer; laxative; antiepileptic; antiparalytic; antirheumatic; treatment of leprosy; treatment of gout; antipyretic; treatment of beriberi, hemiplegia, and eye troubles
Celosia argentia	aphrodisiac, antidiarrheal, treatment of blood disorders and eye diseases
Centratherum anthelminticum[b]	anthelmintic; vermifuge; treatment of urinary diseases; treatment of skin diseases, leprosy, and leukoderma; treatment of nerve and blood disorders; weight reduction
Cissampelos pariera	diuretic, stomachic, antitussive, treatment of urinary troubles, treatment of snakebites, treatment of piles and itching
Clerodendrum inerme	antirheumatic, antipyretic, treatment of lymphoma
Clerodendrum phlomidis	weight reduction, antidiabetic, treatment of measles
Cocculus villosus	antiopiate, refrigerant, antirheumatic, treatment of venereal diseases
Commiphora mukul	stomachic, antiinflammatory, antibacterial in ear infection, antirheumatic, antitussive, antitubercular
Crateva religiosa	kidney stone decalcification, treatment of liver ailments, antibiotic, removes pregnancy marks
Cryptolepis buchanani	cure for rickets, treatment of skin diseases, influences pigmentation, treatment of liver diseases, treatment of lymphoid diseases, blood purifier
Cynodon dactylon	antihemorrhagic, antisyphilitic, tranquilizer, antispasmodic, treatment of insanity, treatment of ophthalmitis, wound healing, promotes conception
Cyperus rotundus	diuretic, emmenagogue, diaphoretic, antipyretic, antibiotic, antiviral
Dalbergia latifolia	blood purifier, treatment of dyspepsia, treatment of leprosy, treatment of obesity, anthelmintic, stomachic, antipyretic

Continued on next page

Table III. *Continued*

Plant	*Uses*
Datura metel	treatment of rabies, antipyretic, antiasthmatic, treatment of skin diseases, treatment of glandular swelling, stimulates brain functions
Dendrocalamus strictus	diuretic, treatment of urinary tract diseases, antimenorrhagic, antidiabetic
Desmodium gangeticum	astringent, antidiarrheal, antipyretic, biliousness, antitussive, antiasthmatic, treatment of snakebite and scorpion sting
Diospyros malabarica	astringent
Echinops echinatus	treatment of gum diseases, facilitates delivery in childbirth, treatment of eye diseases, treatment of dyspepsia, carminative, treatment of skin and blood diseases
Eclipta alba	antidiabetic, influences skin pigmentation, hepatic, treatment of jaundice, antiseptic, treatment of ulcers and wounds
Emblica officinalis	refrigerant, laxative, antihemorrhagic, treatment of jaundice, antiasthmatic, antitussive, treatment of lung inflammation
Erythrina indica	biliousness, expectorant, diminishes central nervous system functions, lactagogue, antidysenteric, anthelmintic
Euphorbia hirata[c]	kidney stone decalcification, antiasthmatic, antitussive, antihistamine, rejuvenator
Ficus bengalensis	antirheumatic, astringent, antidiabetic, poultice for abscesses, antigonorrheal, corrects seminal and ovulation problems, treatment of sore throat
Ficus glomerata	treatment of sores and wounds, treatment of ulcerative colitis, treatment of eye troubles, antimenorrhagic
Ficus hispida	purgative, emetic, treatment of leukoderma/vitiligo
Ficus religiosa	astringent, antigonorrheal, laxative, treatment of hemoptysis, antidiabetic, aphrodisiac, antitussive, alters embryonic sex to male, antimenorrhagic, antiscabietic, treatment of fistula
Gardinia lucida	carminative, treatment of sprue
Gloriosa superba	anticholeraic, treatment of leprosy and skin diseases, antihemorrhoidal, promotes labor pains, treatment of snakebites and scorpion stings
Glycirrhiza glabra	antitussive, eases thirst, demulcent, treatment of influenza, biliousness, treatment of uterine complaints
Grewia tiliaefolia[d]	wound healing, antidysenteric
Gymnea sylvastre	antidiabetic, antiglycosuric, destroys sweet and bitter tastes
Hamilitonia suaveolens	corrects spinal curve
Helicteres isora	demulcent, astringent, treatment of flatulence, antidiabetic, treatment of rabies
Hemidesmus indicus	demulcent, treatment of skin diseases, blood purifier, antisyphilitic, antirheumatic

Table III. *Continued*

Plant	*Uses*
Hiptage bengalensis[e]	antirheumatic, treatment of skin diseases and leprosy, antiasthmatic, insecticide, antiscabietic, reduces waistline
Holarrhena antidysenterica	antidysenteric, astringent, antipyretic, anthelmintic
Indigofera tinctoria	treatment of hydrophobia, epilepsy, and nervous disorders; antitussive; as an ointment; treatment of ulcers; hepatic; treatment of wounds, bites, and burns
Ipomea turpentha	diuretic, demulcent
Jasminum officinale	antitubercular, treatment of eczema, analgesic, improves memory
Jatropha curcas	cleanses eyes, antirheumatic, treatment of paralysis, wound healing
Lannea coromandelica[f]	treatment of rhinitis, shoulder pains, sprains, skin diseases, elephantiasis, ulcers, analgesic, treatment of gallstones and vaginal secretion
Leptadenia reticulata	treatment of night blindness and eye disorders, treatment of pleurisy, wound healing
Loranthus maculatum	induces son's birth
Madhuca indica	astringent, stimulates appetite, antitussive, antihemorrhoidal
Maerua oblongifolia	prevents miscarriage
Mallotus philippensis	anthelmintic, treatment of skin diseases and ulcers
Mangifera indica	treatment of scorpion sting, antihemorrhagic, treatment of lung and intestinal diseases, anthelmintic, treatment of throat diseases
Merremia gangeticum	intestinal vermicide, tranquilizer
Melia azadarach	analgesic, anthelmintic, emmenagogue, treatment of leprosy and smallpox, antirheumatic
Michelia champaka	febrifuge, stimulant, expectorant, astringent, purgative, poultice for abscesses, stomachic, carminative, antigonorrheal
Mimosa pudica	antidiarrheal; antihemorrhagic; treatment of uterus diseases, colitis, and ulcers
Morus alba	antipyretic; treatment of throat ailments, nose bleeds, and leprosy
Mucuna pruriens	aphrodisiac; strengthens vaginal muscles; diuretic; digestion aid; treatment of impotence, scorpion sting, and Parkinson's disease (contains L-dopa)
Nerium indicum	applied externally for cankers and ulcers, reduces swellings, treatment of skin diseases and leprosy, emetic
Nycthanthus arbortristis	antirheumatic, antipyretic, anticholeral, anthelmintic, treatment of colds
Oxalis corniculata	antihemorrhagic, antipyretic, antidysenteric
Phoenix sylvestris	tonic, analgesic, blood purifier, anticholeral
Phyllanthus niruri	antigonorrheal; treatment of genital and urinary tract diseases; treatment of jaundice, swellings, and ulcers

Continued on next page

Table III. *Continued*

Plant	*Uses*
Piper longum	carminative, stimulant, expectorant, antiasthmatic
Plumbego zeylanica	treatment of anorexia, feeding stimulant
Polygonum glabrum	treatment of colic, jaundice, and pneumonia
Psoralea corylifolia	treatment of leukoderma–vitiligo; stomachic; treatment of teeth problems, deafness, eczema, and elephantiasis
Pueraria tuberosa	demulcent, antipyretic, antiinflammatory, tonic, lactagogue, aphrodisiac
Putranjiwa roxburghii	prevents colds, antipyretic, promotes a healthy fetus, treatment of snakebites and erysipelas
Randia dumetorum	treatment of bloody urine and stool, calms stomachache when applied to belly button, aphrodisiac, emetic, antitussive
Rauwolfia serpentiana	sedative; hypnotic; alleviates pain of uterine contractions; treatment of snakebites, insomnia, central nervous system disorders, and high blood pressure
Salix tetrasperma	febrifuge, treatment of rabies and leprosy, antispasmodic
Salvia plebeia	antigonorrheal, antimenorrhagic, antihemorrhagic
Sapindus laurifolius	antispasmodic, treatment of epilepsy, stimulant
Schleichera oleosa	astringent, relief of itching, treatment of ulcers and acne, promotes hair growth
Sida cordifolia	treatment of leprosy, antihemorrhoidal, treatment of facial paralysis and sciatica, antitubercular, heals wounds, aphrodisiac, antigonorrheal, cardiac muscle and nerve tonic, removes mouth dryness due to alcohol consumption
Solanum indicum	carminative, expectorant, antiasthmatic, analgesic, antipyretic, relief of itching, antiinflammatory
Solanum nigrum	antipyretic, treatment of eye diseases and hydrophobia, cathartic, diuretic
Solanum xanthocarpum	antitussive, treatment of colic, antiasthmatic, antirheumatic, induces maleness to embryo, treatment of eye diseases
Stereospermum chelonoides	treatment of nerve disorders
Strychonus nux-vomica	digestion aid, treatment of nerve disorders, tonic, stimulant, treatment of skin diseases and leukoderma, termite and fish poison
Syzygium cumini	astringent, antidiarrheal in children, stomachic, antibilious, antidiabetic
Tamarindus indica	refrigerant; digestion aid; carminative; treatment of febrile diseases, constipation, and intoxication
Tecomella undulatea	bone setting, thins blood clots, rejuvenation, antiviral, antitumor
Tephrosia hirata	treatment of spleen and skin diseases
Terminalia arjuna	used for cardiac rhythm of muscles, Ca^{+-} dependent function

Table III. *Continued*

Plant	*Uses*
Terminalia belerica	astringent, tonic, laxative, antihemorrhoidal, diuretic, treatment of leprosy, digestion aid, analgesic, narcotic, antitussive, treatment of sore throat
Tinospora cordifolia	stomachic, antipyretic, aphrodisiac, antigonorrheal, antimalarial, lactogogue, treatment of leprosy
Tribulus terrestris	rejuvenation, antiinflammatory, diuretic, treatment of kidney disorders
Trichosanthes bracteata	hydragogue, cathartic, antiasthmatic, applied externally in hemicrania and ozoena, treatment of carbuncles, analgesic, treatment of lung diseases in cattle, antiviral, treatment of herpes
Verononia cinerea	antipyretic, kidney stone decalcification, induces sleep
Viscum articulata[g]	treatment of cardiac properties akin to digitalis, antiergotic
Vitex negundo	vermifuge, analgesic, antirheumatic, antihistamine, antiscrofulous, treatment of enlarged spleen, antitubercular
Withania somnifera	antiinflammatory; antitumor; treatment of lumbago, sciatica, and arthritis; antiasthmatic
Woodfordia fruiticosa	antidiarrheal, antimenorrhagic, treatment of vaginal bleeding
Zizyphus oenoplea	astringent, antidiarrheal, treatment of throat diseases and cancer

NOTE: The plants were collected Nov.-Dec. 1983 from the Gir Forest and Mount Abu; selected therapeutic uses are listed *(17, 41-49)*.
[a]*Amaranthus lividus.* [b]Synonym *Veronia anthelmintica.* [c]Synonym *Euphorbia pilulifera.* [d]*Grewia asiatica* and *Grewia obatera.* [e]Synonym *Hiptage madablota.* [f]Other species *Lannea pinnatifeda* and *Lannea grandis.* [g]Synonym *Viscum articulata.*

Table IV. Medications and Ingredients for Leukoderma Treatment

System	*Medications*
Ayurvedic	internal and topical oils
	Psoralea corylifolia, bakuchi, plant
	Ficus hispida, vad, plant
	tamra (copper) bhasma
	malla (arsenic) bhasma
Siddha	*Strychnos nux-vomica,* etti, copper plant
	Vishamuthi thailam, oil
	copper, gold, and iron bhasma
Unani-tibb	*Ammi majus,* aatrilal, plant
Allopathic	psoralen, synthetic, chemical

vitiligo patients *(53)* and may reflect the failure of copper to remain tissue bound.

Many components of these ancient traditional treatments are consistent with modern concepts of melanin synthesis, but the problem of "treatment to cure" leukoderma is much more complex *(35)*. One must consider many other aspects of this disease: the anatomical and structural features of the skin localizing the site of production of melanocytes, the migration of melanocytes, the basal cellular layer where gene control of pigmentation initiates, the biochemical pathway of melanin synthesis, and the chemical constituents of medication and their role in achieving repigmentation.

Psoralen is a known component of many plant species *(54, 55)*. Psoralen has been used with some success to treat vitiligo [Stolar *(31)*; also see the comments of Fitzpatrick that claim the good results may be exaggerated]. The ancient traditional treatments seem to be more efficacious *(26, 56)*, possibly because the latter are multicomponent and hence may provide the necessary components for the numerous steps required for melanization.

8-Methoxypsoralen and other analogues are used primarily in "suntan pills" and suntan screens. At higher dosages 8-methoxypsoralen may induce skin cancer, but at lower concentrations in the diet 8-methoxypsoralen reduces skin cancer *(50, 57)*. One hour following oral ingestion of psoralen, the skin sensitivity to UV light increases and reaches its peak in 2 h. Skin sensitivity disappears in 8 h. The photoreaction seems to enhance melanization. Psoralen is observed to increase the thickness of the stratum corneum of the epidermis, the horny layer, thus providing a screen. The precise biochemical interaction in melanization is not known. Psoralen may be involved in mobilization rather than synthesis of pigments. Psoralen is known to accumulate in melanocytes *(50)*.

Crude acetone–water extracts of *P. corylifolia* and several species of *Ficus* and 8-methoxypsoralen nick *Escherichia coli* DNA upon UV irradiation (N. G. Patel and Y.-C. Tse-Dinh, personal communication). This reaction seems to occur in three steps: intercalation of psoralen in DNA, formation of photoadducts, and fragmentation of DNA. The details of the photoadduct mechanism were investigated *(58)* recently and reviewed *(59)* in detail. The mechanism occurs by a two-step cycloaddition of nucleosides, first to the 4′,5′ (furan) side and then to the 3,4 (pyrone) side of the psoralen molecule. The significance of this in vitro mechanism to in vivo pigmentation is not clear. Plants contain both linear (psoralen type) and angular (angelin type) analogues, which form bis- and monophotoadducts, respectively. What role either individually or in combination these molecules play in ayurvedic preparations is unknown.

Collectively, the following facts emerge.

- Leukoderma is curable, though with difficulty and not with 100% success. Better understanding of the interacting mechanisms could lead us to a complete cure.
- The plant species frequently used are *P. corylifolia, F. hispida* (other *Ficus* spp. also), and *A. majus*.
- These plant species all contain psoralen, which plays an important role in melanization. When administered alone, psoralen seems to be less efficacious and has side effects compared to complex mixtures of traditional medicines.
- Psoralen (as suntan pills) induces hyperpigmentation over the entire body, but the traditional treatments seem to be selective and induce pigmentation primarily in leukodermal regions, though in isolated cases increased pigmentation in other regions is also observed.
- The medication is administered both orally and topically.
- Sunlight (UV) is essential for repigmentation.
- Copper increases the efficacy, when either taken as bhasmas or obtained from copper plants. Investigation of the effects of copper–psoralen formulations and determination of the possible lowering of the required dose, toxicity, and side effects would be interesting.

This review is restricted to only a few aspects of leukoderma. Other interesting areas for future studies include the following:

- Biochemical differences in normal and depigmented cells.
- Genetic aspects.
- Basic gene expression mechanism to reveal whether depigmented cells lack production of crucial enzymes, such as tyrosinase and polymerase.
- The mechanism of the spread of depigmentation is intriguing: Do the depigmented cells produce inhibitory metabolites, such as phenylalanine, ascorbic acid, or other reducing compounds?

- The role of MSH, ACTH, and other hormones.
- The significance of dietary restrictions to ayurvedic medication.
- The contribution of etiological factors. An inquiry as to why leukoderma is absent in rural areas could be valuable.

Apparently, without the knowledge of precise molecular mechanisms, the ancient traditional medicine seems to have achieved significant success in treating leukoderma.

Ayurveda transcends from a simple common sense preventive approach to extremely complex psychosomatic and belief components of human health care. In this realm, ayurveda could provide exciting and potentially rewarding opportunities to scientists trained in physical, chemical, biological, and social sciences.

ACKNOWLEDGMENTS

In part, the reported information was collected during the Ayurvedic Medicinal Plants: India expedition, 1983, sponsored by Earthwatch, Belmont, Mass.; by Narayan G. Patel, principal investigator; and a group of volunteers from the United States and India. I sincerely thank the Earthwatch expedition volunteers, who selflessly assisted in the collection of 2000 plants from the hills of Mount Abu and the Gir Forest of western India in 1983. I am thankful to Earthwatch for providing volunteers. Many volunteers and Indian scientists who graciously hosted us and helped us collect plants deserve our sincere appreciation. My special thanks go to R. D. Adatia for the taxonomic identification and to Dan Nicolson for taxonomic nomenclature; and to Steven Skopik, David Boykin, and William Cain for their comments and corrections of the manuscript. I respectfully extend my heartfelt gratitude in dedicating this chapter to Shri Bapalal Vaidya, the revered and dedicated ayurvedic doctor, who passed away at the age of 88 on the day the expedition ended. He shared with us his encyclopedic knowledge of Indian medicinal plants every moment of the 6-week expedition.

LITERATURE CITED

1. Dash, B. In "Scientists"; Raghavan, V., Ed.; Ministry of Information and Broadcasting, Government of India: New Delhi, 1979.

2. Shashtri, G. M. "Charak Samhita"; Sastu Sahitya Vardhak Karyalaya: Ahmedabad, India; Vols. 1-5.
3. Sankaran, P. S.; Deshapande, P. J. In "Scientists"; Raghavan, V., Ed.; Ministry of Information and Broadcasting, Government of India: New Delhi, 1979.
4. Shastri, K. G. "Sushrut Ayurveda"; Sastu Sahitya Vardhak Karyalaya: Ahmedabad, India, 1973; Vols. 1-2.
5. Leslie, C. "Asian Medical Systems: A Comparative Study"; Univ. of California Press: Berkeley, 1979.
6. Dash, B.; Kashyap, L. "Diagnosis and Treatment of Diseases in Ayurveda"; Concept Publishing Company: New Delhi, 1982; Vols. 1-2.
7. Lad, V. "Ayurveda: The Science of Self Healing"; Lotus: Santa Fe, N. Mex., 1974.
8. Thakkur, G. G. "Ayurveda: The Indian Art and Science of Medicine"; ASI: New York, 1981.
9. Parikh, R. J. "Madhavnidan"; Sastu Sahitya Karyalaya: Ahmedabad, India, 1975; pp. 496-98.
10. Bajaj, T. S. "Insulin and Metabolism"; Elsevier: Amsterdam, 1977.
11. Brodoff, B. N.; Bleicher, S. J. "Diabetes Mellitus and Obesity"; Williams and Wilkins: Baltimore, 1982.
12. Melchionda, N.; Horwitz, D. L.; Schade, D. S., Eds. "Recent Advances in Obesity and Diabetes Research"; Raven: New York, 1984; p. 414.
13. Travis, R. H.; Sayers, C. In "Pharmacological Basis of Therapeutics"; Goodman, L. S.; Gilman, A., Eds.; McMillian: London, 1970.
14. Campbell, G. C. "Oral Hypoglycemic Agents"; Academic: New York, 1969; p. 482.
15. Murthy, N. A.; Randley, D. P. "Ayurvedic Cure for Common Diseases"; Orient Paperbacks: New Delhi, 1982; pp. 191-94.
16. Patel, N. G., unpublished data, 1983.
17. "The Ayurvedic Formulary of India"; Government of India: New Delhi, 1978; Part 1.
18. Barot, K. C.; Gupta, P. C.; Deshpande, I. S.; Agravat, S. B.; Suthar, A. K. "A Comparative Study of Jambu and Mamejjak as Hypoglycemic Agents in Diabetes Mellitus," Ayurvedic Research Seminar, Gujarat, Ayurveda Univ., Jamnagar, India, 1976, pp. 65-69.
19. Chaturvedi, G. N.; Subramaniyam, P. R.; Tiwari, S. K.; Singh, K. P. *Ancient Sci. Life* **1983,** *3(4),* 216-24.
20. Chaudhury, R. R.; Vohra, S. B. In "Advances in Research in Indian Medicine"; Uidupa, K. N.; Chaturvedi, G. N.; Tripathi, S. N., Eds.; Banaras Hindu Univ.: Varanasi, India, 1970; pp. 57-75.
21. Pillai, N. R.; Ghosh, D.; Uma, R.; Kumar, A. *Bull. Med. Ethno. Bot. Res.* **1980,** *1(2),* 234-42.
22. Mehta, B. N. In "Rasatantrasar One Siddhaprayogsangrah"; Krishnagopal Ayurved Bhavan: Ajmer, India, 1967; pp. 84-243.
23. Patel, N. G.; Dixit, P. K. *Nature* **1964,** *202,* 189-90.
24. Riley, V.; Fortner, J. G. *Ann. NY Acad. Sci.* **1963,** *100,* 1-1123.
25. Desai, V. S.; Pandya, K. C.; Dave, G. K.; Hirapara, P., personal communication, 1980.
26. Razzack, M. A. *Proc. Sem. Bars (Leukoderma),* Central Council for Research in Unani Medicine, Hyderabad, India, 1979.
27. Shah, V. C. "Genetic, Biochemical and Cytological Studies on Leukoderma"; Gujarat Univ. Publication, Ahmedabad, India, 1982, p. 72.

28. Mitra, R.; Pandey, H. C. *Proc. Sem. Bars (Leukoderma),* Central Council for Research in Unani Medicine, Hyderabad, India, 1979, pp. 132–36.
29. Mukherjee, G. D. *Proc. Sem. Bars (Vitiligo),* Central Council for Research in Unani Medicine, Hyderabad, India, 1979, pp. 97–100.
30. Zafarullah, M.; Vohra, S. B. *Proc. Sem. Bars (Vitiligo),* Central Council for Research in Unani Medicine, Hyderabad, India, 1979, pp. 125–31.
31. Stolar, R. *Ann. NY Acad. Sci.* **1963,** *100,* 58–75.
32. Taiyab, H. M.; Husain, S. M. S.; Salahuddin; Sultana, A. *Proc. Sem. Bars (Vitiligo),* Central Council for Research in Unani Medicine, Hyderabad, India, 1979, pp. 101–12.
33. Khan, M. M.; Mirza, M. A. *Proc. Sem. Bars (Vitiligo),* Central Council for Research in Unani Medicine, Hyderabad, India, 1979, pp. 35–40.
34. Punshi, S. K. *Q. Med. Rev.* **1979,** *30(4),* 1–46.
35. Nordlund, J. J.; Lerner, A. B. *Arch. Dermatol.* **1982,** *118,* 5–8.
36. Iqbal, A. M.; Ali Khan, M. M.; Muneer, A.; Mastan, A. *Proc. Sem. Bars (Leukoderma),* Central Council for Research in Unani Medicine, Hyderabad, India, 1979, pp.9–15.
37. Seiji, M.; Shimao, K.; Birbeck, M. S. C.; Fitzpatrick, T. B. *Ann. NY Acad. Sci.* **1963,** *100,* 497–533.
38. Nanal, B. P.; Ranade, S. S. "Study of Kokodumber in the Treatment of Shwitra"; Ayurvedic Research Seminar, Gujarat Ayurveda Univ., Jamnagar, India, 1976, pp. 24–31.
39. Trivedi, V. P.; Ansari, Z.; Shukla, K. *Proc. Sem. Bars (Vitiligo),* Central Council for Research in Unani Medicine; Hyderabad, India, 1977, pp. 68–77.
40. Jopat, P. D.; Karnick, C. R. *Proc. Sem. Bars (Vitiligo),* Central Council for Research in Unani Medicine, Hyderabad, India, 1979, pp. 83–88.
41. Chopra, R. N.; Chopra, I.C.; Handa, K. L.; Kapur, L.D. "Indigenous Drugs of India"; Academic: Calcutta, India, 1982; p. 816.
42. Dash, B.; Kashyap, L. "Materia Medica of Ayurveda"; Concept Publishing Company: New Delhi, India, 1980; Vols. 1–2.
43. Kirtikar, K. R.; Basu, B. D. "Indian Medicinal Plants"; Bishen Sing Mahendra Pal Singh: Dehradun, India, 1980; Vols. 1–4.
44. Labadie, R. P. *Proc. Int. Workshop Priorities Study of Indian Med.* **1984,** Meulenbeld, G. J., Ed., pp. 209–22.
45. Nandkarni, K. M. "Indian Material Medica"; Popular Prakashan Private Ltd.: Bombay, India, 1982; Vols. 1–2.
46. Shastri Pade, S. D. "Aryabhishak"; translated by Yyas, H. B.; Sastu Sahitya Vardhak Karyalaya: Ahmedabad, India, 1976.
47. Vaidya, B. G. "Nighantu Adars"; Shri Swami Atmanand Saraswati, Ayurvedic Sahakari Pharmacy Ltd.: Surat, India, 1965; Vols. 1–2.
48. Vaidya, B. G. "Some Controversial Drugs in Indian Medicine"; Chaukhambha Orientalia: Varanasi, India, 1982; p. 671.
49. Iyengar, M. A. "Biolography of Investigated Indian Medicinal Plants 1950-75"; Manipal Power: Manipal, India, 1976.
50. Becker, S. W. *J. Am. Med. Assoc.* **1960,** *173,* 1483–85.
51. Kumaraswamy, R.; Maria, J. A.; Meenakshi, N. E.; Gnana Dass, D. *Proc. Sem. Bars (Vitiligo),* Central Council for Research in Unani Medicine, Hyderabad, India, 1979, pp. 169–76.
52. Sharma, T. N.; Joshi, D.; Sen, S. P. *J. Res. Indian Med.* **1966,** *1,* 78–80.
53. Tehera, S. S.; Quamaruddin, S. *Proc. Sem. Bars (Vitiligo)* Central Council for Research in Unani Medicine, Hyderabad, India, 1979, pp. 165–68.

54. Fowlks, W. L. *J. Invest. Dermatol.* **1959,** *32,* 249–54.
55. Pathak, M.A.; Farrington, D.; Fitzpatrick, T. B. *J. Invest. Dermatol.* **1962,** 225–36.
56. Pandya, K. C.; Dave, G. K. "Study of B-yoga in the Treatment of Leukoderma/Vitiligo," All India Ayurveda Shastra Charcha Parishad, 1978, pp. 35–43.
57. Farber, E. M.; Abel, E. A.; Cox, A. J. *Arch. Dermatol.* **1982,** *119,* 426–31.
58. Kane, D.; Straub, K.; Hearst, J. E.; Rapoport, H. *J. Am. Chem. Soc.* **1982,** *104(24),* 6754–64.
59. Pearlman, D. A.; Holbrook, S. R.; Pirkle, D. H.; Kim, S. H. *Science* **1985,** *227,* 1304–15.

0939-1/86/0065$07.00/0

Rhizophora mangle (Reproduced with permission. Copyright 1981 Reference Publications, Inc.)

Chapter 4

Fijian Medicinal Plants

R. C. CAMBIE*

Fiji is unlike some countries where the introduction and continuing demand for ethical drugs has led to a decline in traditional medicinal practices. With a more open approach by health authorities toward traditional medicines, a revival of knowledge of medicinal plants has occurred.

Of approximately 2300 species of plants to be found in Fiji, some 450 have been recorded as having been used for medicinal purposes. Many of the latter, such as *Cordyline terminalis* (family Agavaceae), *Ageratum conyzoides* (Asteraceae), *Hibiscus tiliaceus* (Malvaceae), and *Centella asiatica* (Umbelliferae), are of pantropic distribution and/or are of Indian origin. Also, many Fijian medicinal plants are common weeds or coastal plants of the type found around villages, and probably 75% of the plants used medicinally are readily available. Medicinal uses are reported for plants in all of the following categories: endemic, indigenous, early introductions, and recent introductions. Medicinal uses for the endemic species are restricted to Fiji, but several of the widespread indigenous species and most introduced species are utilized elsewhere. Indeed, some of the introduced species were brought to Fiji for their medicinal properties. Certain families, such as the Euphorbiaceae and Rubiaceae, are important medicinally, but other families are scarcely represented, despite the presence of

*Department of Chemistry
University of Auckland
Auckland, New Zealand

many of their species in Fiji. Conifers, mosses, algae, lichens, and fungi are not used medicinally.

Mystical aspects play a large part in the medicinal properties of Fijian plants. According to tradition, two kinds of medicine *(ne i drotu)* exist, the medicines of death and the medicines of life. Paramount among both of these medicines are herbal or plant remedies—many of these remedies are of pathological value only, and others are only autosuggestive cures. For example, leaves of beneviriviri *(Jasminum betchei)* were chewed when Fijians were in a strange district to prevent their being bewitched *(vakadrauni-kaued)*. In a similar fashion, the leaves of kalabuci *(Acalypha insulana)* were chewed to ensure safety when traveling on unfamiliar tracks, and those of kalabuci damu *(Acalypha wilkesi-ana)* were chewed to keep one safe from all harm.

A number of trees were treated with what amounted to superstitious veneration. Examples of such sacred or devil trees include dawa *(Pometia pinnata),* merikula *(Maesa persicifolia),* nokonoko *(Casuarina equisetifolia),* tarawau *(Dracontomelon vitiense),* baka *(Ficus obliqua),* and koukoutangane *(Dipteris conjugata)*. Boia *(Alpinia boia)* was under the care of the *veli* (spirits of the woods), and it was believed that anyone who touched it would be overtaken by some evil. To cut boia or to cut vesi *(Intsia bijuga)* was to risk even worse calamities. Bovo *(Mussaenda frondosa)* was included among the mystery plants of Fiji, and in early days bovo came into evil repute. Early Fijians believed the *veli* made their yaqona (or kava), a popular beverage, from the roots of yaqoyaqona *(Piper puberulum)* and thus the plant was sacred. The moli karokaro *(Citrus limon),* common lemons, are said to be grown near houses not only for their fruits but also as a protection against evil. *Tevoras* (malign influences) could then work no harm, since the moli karokaro wielded a more potent charm.

Most of the present-day diseases to be found in Fiji were unknown in aboriginal times and have been introduced to the islands by Europeans. At various times the local population of Fiji has been ravaged by epidemics, including measles, influenza, cholera, and venereal disease. For example, the lila (measles) epidemic of 1875 is estimated to have wiped out some 40,000 individuals out of a population of about 150,000; it was described by one author as "Fiji's darkest hour". Another measles epidemic in 1903 is reported to have attacked nearly half of the Fijian population, and an influenza epidemic of 1918 attacked at least 80% of the population and resulted in the death of approximately 5% of the population. Because of the early unavailability of European drugs, many plant remedies were tried, and these plants were often the nearest species found at hand.

Little chemical investigation has been carried out on endemic species, although interest in this area has increased as a result of activity at the University of the South Pacific. On the other hand, some of the pantropic species, especially those of Indian origin, have been examined chemically in an extensive manner. In the following discussion, the chemical constituents of some Fijian species are examined and, where possible, are related to their reported medicinal uses.

Cananga odorata (Lam.) Hook. f. & Thoms. Family Annonaceae (Fijian: Makosi)

Cananga odorata is a medium to large tree, originally from southern Asia, which is cultivated and also extensively naturalized in Fiji. The tree provides a useful timber, and the pale yellow flowers are used to scent coconut oil and for the preparation of leis.

The leaves are reputed to be a sure remedy for ophthalalmia or "eye blight", a painful irritation of the eyelids that, if chronic, can lead to permanent loss of sight. A decoction of the leaves is used for boils in the ear, and a decoction of the bark is used for stomach muscular pains and for menstrual disorders. The bark is also reported to be good as an emmenagogue and to be antisyphilitic. The stem of young plants is reputed to be part of an external remedy for back pains, and the root is said to be used for cancer. The crushed flowers are also used in Pacific folk medicine.

The essential oil contains pinene, limonene, phellandrene, cineole, (-)-linalool, geraniol, benzyl alcohol, *p*-cresol, eugenol, isoeugenol, methyleugenol, benzoic and salicylic acid esters, sesquiterpenes such as caryophyllene and cadinene, and an alkaloid. The plant also contains cyanogenic material. The essential oil of the flowers that has been examined in Fiji contains linalool (52%), benzyl benzoate (12%), and farnesol (21%) as the only major constituents. The plant also contains the alkaloid eupolauridine. Eugenol has been used for the treatment of peptic ulcers and has been used internally as an antiseptic and an antipyretic. Eugenol is employed externally by dentists as an antiseptic and as a feeble anesthetic. Benzyl alcohol is an antiseptic, an antipruritic, and a local anesthetic, and benzyl benzoate is used for scabies and pediculosis and has been used in the past as a muscle relaxant. Linalool has strong germicidal properties.

Alstonia vitiensis Seem.
Family Apocynaceae (Fijian: Sorua)

Alstonia vitiensis is a medium-sized endemic tree that gives a copious white latex.

A decoction of the leaves is reported to be used as a tonic, and supposedly a very painful eye disease known locally as oika can be cured by pounding the bark and injecting the juice into the eyes.

The plant (var. *novo ebudica monachino)* contains the indole alkaloids pleiocarpamine, alstovine, vincorine, cabucraline, and quaternoxine, and probably the alkaloids are responsible for the tonic properties of the leaves.

Bleekeria vitiensis A.C. Sm.
Family Apocynaceae (Fijian: Dogodogo)

Bleekeria vitiensis (see Figure 1) is a moderately common small endemic tree that is reported to have anticancer properties.

Most parts of the plant contain alkaloids including the bark, which is the best known source of the antitumor alkaloid 9-methoxyellipticine.

N-Methyl-9-hydroxyellipticinium acetate also possesses antitumor activity and is currently under clinical trial in humans. The intercalation property of the latter compound with DNA is probably not sufficient to explain its antimitotic activity, and a recent study suggests that its activity may be related to its ease of formation of ketalic nucleoside adducts.

Cerbera manghas L.
Family Apocynaceae (Fijian: Vasa)

Cerbera manghas is a common softwood tree whose white fragrant tubular flowers are used for leis.

The tips of the leaves are steeped in coconut oil, which is then used as an ointment for skin diseases such as scabies—reputably,

Figure 1. Bleekeria vitiensis *A.C. Sm. (family Apocynaceae) (Fijian: Dogodogo).*

this ointment can be kept for several years. The leaves and fruit are a powerful emetic, and the root is used as a cathartic. The root and bark provide a powerful purgative and are used for constipation, and decoctions of the bark are used for liver disorders and as an abortifacient.

The plant contains cardiac glycosides, including cerberoside, therobioside, thevetin β, deacetyltanghinin, and neriifolin. Hydrolysis of extracts of the twigs and stem bark gives the pseudoindican plumieride. The seeds contain the glycoside cerberin and the bitter principle odollin. Cerberin is a heart poison, yielding isocerberin, digitonide, cerberose, and cerberigenin.

Mikania micrantha H.B. and K. (syn. *Mikania scandens* Willd.) Family Asteraceae (Fijian: Wabosucu)

Mikania micrantha is a very common introduced climbing weed that earns the name of "mile a minute" on account of its amazing rapidity of growth.

The plant is attributed with a number of medicinal properties, but in particular, the leaves are used externally to stop bleeding and to act as an antiseptic. The leaves are used as a remedy for skin irritations such as bee stings. A decoction is used internally for high blood pressure, diabetes, and stomachache and is also used to hasten birth for expectant mothers who are over their normal term of pregnancy. The plant is said to be active against cancer.

M. micrantha contains the sesquiterpene lactone mikanolide, which may or may not be responsible for some of the physiological properties but is almost certainly responsible for the anticancer activity.

Wedelia biflora (L.) DC.
Family Asteraceae (Fijian: Kovekove)

Wedelia biflora is a branching perennial shrub that is distributed from Malaysia and southern Asia to Polynesia.

The leaves are soaked in coconut oil and used to massage sprained or bruised limbs, and a decoction of the leaves is used for bacillary dysentery, infective hepatitis, hemorrhoids, and an infected bladder. The leaves and stems are used for treating appendicitis and eczyma and the stems for pimples. The leaves are used with those of other species for muscular spasms, convulsions, and stomachache. The bark is used with the milk of coconut and the root of masi (a *Ficus* sp.) for treating fish poisoning.

The essential oil of the leaves contains (–)-α-pinene as the major component. Other *Wedelia* sp. contain diterpenoids, especially of the kaurene acid type, and *Wedelia asperrima* contains wedeloside, which has antitumor activity.

Hoya australis R. Br.
Family Asclepiadaceae (Fijian: Draubibi)

Hoya australis is a common climbing vine that is characterized by its sweetly scented waxy flowers with red centers (Figure 2). A decoction of the leaves is used for convulsions and as a tonic for consumption, rheumatism, and cancerous growths. When mixed with vativati *(Microsorium scolopendria),* the decoction provides a drink that is used to strengthen mothers after childbirth and for assisting in postnatal discharge.

Figure 2. Hoya australis *R. Br. (family Asclepiadaceae) (Fijian: Draubibi).*

The plant contains a series of triterpenes in the latex, and the leaves also contain large amounts of chlorogenic acid, isochlorogenic acid, apigenin-type flavones, and phenolic depsides, for example, *p*-coumaric acid esters, but none of these compounds have obvious links with the reputed medicinal properties of the plant.

Cassia alata L. Family Caesalpiniaceae (Fijian: Kadrala)

Cassia alata is a common introduced ornamental shrub that is a native of tropical America but is now pantropic. The plant has brilliant yellow candlestick flower spikes from whence the local name "Roman candle tree" arises (Figure 3).

The leaves or seeds are crushed, and the juice is rubbed on the afflicted part to cure parasitical skin diseases, especially ringworm. The plant is also used in India and in many Southeast Asian and Pacific countries for the same purpose.

Figure 3. Cassia alata *L. (family Caesalpiniaceae) (Fijian: Kadrala).*

The active ingredients are almost certainly its anthraquinones, which include emodin, aloe emodin (plus glucoside), rhein (plus glucoside), chrysophanol, isochrysophanol, and physcion 1-glucoside. The plant also contains cassiaxanthone.

Compounds such as emodin, rhein, and chrysophanol are all cathartics and, according to Smythies, should all share weak tumor-promoting action.

Intsia bijuga O. Kuntze
Family Caesalpiniaceae (Fijian: Vesi)

Intsia bijuga is a common spreading tree that was formerly a sacred tree. The timber had extensive uses for canoes, yaqona bowls, and clubs.

A decoction of the bark is used for rheumatism, chills, diarrhea, rheumatoid arthritis, and muscular rigidity. A decoction of the leaves was drunk when the body supposedly had been possessed by a spirit—a magical rather than a medicinal use. The leaves were used with the roots of weleti *(Carica papaya)* for toothache.

The leaves contain traces of alkaloids, the bark contains tannins, and the heartwood contains flavonoids (for example, naringenin, myricetin, and robinetin), stilbenes, and water-soluble polymers. However, other than for the tannins, no obvious linkup with the medicinal properties is observed. Flavonoids such as naringenin have nutritional significance because they may influence the metabolism of ascorbic acid. They also have capillary resistance, where the influence is pharmacological rather than nutritional.

Calophyllum inophyllum L.
Family Clusiaceae (Fijian: Dilo)

Calophyllum inophyllum is a large spreading coastal timber tree of wide distribution that is assumed to be an aboriginal introduction. The tree produces large round fruits with a bittersweet kernel that affords a resinous greenish oil; in 1870 the oil commanded a ready sale on the European market. When broken, the twigs exude a white milky sap.

The oil from the fruit is greatly valued as a liniment by those who suffer from rheumatism, pains in the joints, and bruises. The oil is also applied to suppurating wounds including coral sores, and the oil is said to be commonly used for rubbing on the limbs of children who are slow at learning to walk. The root and bark are also used for rheumatic pains. The juice from the crushed leaves is said to relieve longstanding irritation of the eyes or conjuctivitis. The green fruit is reputed to be used for tuberculosis, and the bark with those of other species is said to be used for treating infected teeth.

Extensive chemical work has been carried out on the seed, which yields an essential oil of unknown composition and a fixed oil composed of the esters of oleic, linoleic, and other fatty acids. The resinous fraction of the seed oil contains several piscicidal phenylcoumarins, for example, ponnalide, calophyllolide, inophyllolide, and the derived calophyllic and inophyllic acids. The latter also occurs in the bark. Extracts can stimulate phagocytosis by the reticuloendothelial systems as determined by protection of mice against a lethal dose of *Escherichia coli.*

Aleurites moluccana (L.) Willd.
Family Euphorbiaceae (Fijian: Lauci)

Aleurites moluccana is a large, handsome tree with conspicuous whitish palmate leaves that is common in Fiji and other Pacific Islands. The ground beneath the tree is often covered with its large hard black nuts. The seeds contain an oil that was formerly used for polishing wood and as a dye in tattooing; the oil was included in the London Exhibition of 1862. It is sometimes referred to as the "candlenut tree" as the seeds were used as a string to provide light like a candle.

The leaves, bark, and fruit are used for a diverse range of medicinal purposes, which include their use as a tonic and as a purgative. The oil of the seeds (called "Tongan oil" in Tonga) is said to be particularly purgative.

The oil of the kernel contains esters of linolenic acid and 9,14-dihydroxy-10,12-octadecadienoic acid. Both immature and mature fruits give positive tests for alkaloids, and probably the alkaloids are responsible for the tonic properties.

Excoecaria agallocha L.
Family Euphorbiaceae (Fijian: Sinugaga)

Excoecaria agallocha is a common coastal tree that is found in both Southeast Asia and Polynesia. The tree yields a copious acrid white latex that causes an intense irritation if it comes into contact with the skin, and reputedly the latex may cause blindness if rubbed into the eyes.

According to Seemann, who was the first to study the flora of Fiji, the early Fijians used to place leprosy sufferers in an empty house in which they would light a small fire on which pieces of sinugaga wood would be placed. The leper would be bound and suspended upside down over the fire in the midst of the suffocating smoke. The smoke caused intense pain, and although it is reputed to have cured, it probably killed some.

The leaves are used to make a decoction that is used for sore eyes and are chewed for "unusual" pain in the eyes and to provide relief from a sore throat and headache.

The bark contains tannins; the latex contains mannitol, behenic acid, and unidentified alcohols; and the wood latex contains the triterpenes β-amyrin, β-amyrenone, 3β-epiamyrin, cycloartenol, and unidentified glycerides of C_{24}–C_{32} acids. A piscicidal component used for catching fish in New Caledonia has also been isolated.

Scaevola taccada Roxb.
Family Goodeniaceae (Fijian: Vevedu)

Scaevola taccada is a thick-stemmed shrub that is common on seashores. The small white flowers are unusual in that they appear to have the petals missing from the upper side.

A decoction of the root is used for stomachache, while a decoction of the bark and leaves is used for a relapse from an illness. The plant is used extensively for medicinal purposes in Southeast Asia and Polynesia.

The plant is said to contain a bitter principle scaevolin and two unidentified glycosides, and the leaves contain chlorogenic acid and give positive tests for saponins.

Gyrocarpus americanus Jacq.
(syn. *Gyrocarpus jacquinii* Roxb.)
Family Gyrocarpaceae (Fijian: Wiriwiri)

A decoction of the leaves of this large tree is used for epilepsy, for bronchial troubles, and especially for women who have just given birth. A decoction of the scraped bark or the leaves is used as a remedy for constipation, and the decoction is said to be a strong purgative and to be used as a remedy for dysentery. The bark is also used for treating arthritis and rheumatism.

The bark contains the alkaloids phaeanthine and (+)-magnocurarine, which have curare-like activity and thus could find use in the treatment of certain heart diseases. The bark and leaves contain *O*-desmethylphaeanthine and the leaves contain phaeanthine.

Hernandia nymphaeifolia
(syn. *Hernandia peltata)*
Family Hernandiaceae (Fijian: Evuevu)

Hernandia nymphaeifolia is a common large spreading coastal tree or shrub that is found from East Africa to Pitcairn Island.

A decoction of the leaves is used to treat menstrual disorders and to ease childbirth. The bark is reputed to have reliable contraceptive properties and to be used as a tonic after childbirth, for urinary infections, and for menstrual pain.

The bark contains alkaloids and lignins, for example, peltatin, epiaschantin, and epieudesmin. Hydrolysis of the leaves gives the flavonoids quercetin and kaempferol.

Cinnamomum pedatinervium Meisn.
Family Lauraceae (Fijian: Macou)

Cinnamomum pedatinervium is a handsome endemic tree with aromatic leaves. The bark has a fine aromatic odor and is used for scenting coconut oil.

The plant is used as a sudorific, that is, it produces copious perspiration. The bark contains terpenes; linalool; aldehydes; esters; and the aromatic compounds eugenol, eugenol methyl ether, and safrole.

Eugenol has been employed in peptic ulcer treatment and formerly was used internally as an antiseptic and antipyretic. Eugenol was also employed externally by dentists as an antiseptic and feeble anesthetic. Safrole is used in perfumery but also has antiseptic, pediculicide, and carminative properties.

Hibiscus tiliaceus Borss.
Family Malvaceae (Fijian: Vau)

Hibiscus tiliaceus is a very common small spreading tree that has large yellow flowers with a brown or deep maroon center (Figure 4).

The leaves are wrapped on bone fractures and sprains, and the stem is said to be part of an internal remedy for treating ulcers and for internal injury. The leaves are also used for amoebic dysentery, for infected wounds, for fish poisoning, as a tonic for mothers after childbirth, and with the leaves of other species for the relief of menstrual pain. The leaves are also said to have antifertility properties, but this statement has not been authenticated. The sap obtained by pounding the inner bark is used as an ointment for treating scabies.

The wood collected from Fiji contains a series of sesquiterpenoid quinones, hibiscones A–D, which may account for some of

Figure 4. Hibiscus tiliaceus *Borss. (family Malvaceae) (Fijian: Vau).*

their pharmacological activity. The wood also contains lapachol, while the flowers contain gossypetin, kaempferol, and quercetin.

Entada phaseolides (L.) Merr. Family Mimosaceae (Fijian: Walai)

Entada phaseolides is a creeper or liana with long tendrils that is common in mangrove swamps and seashore areas.

The leaves have been used for a multitude of medicinal purposes, for example, as a remedy for filariasis or elephantiasis; as a tonic against cancer, venereal disease, thrush, and hemorrhoids; and with the leaves of other species to treat convulsions, to shorten the period of labor of expectant mothers, and to assist in the expelling of the placenta after childbirth.

The leaves contain oleanolic acid and saponins, which have a strong hemolytic action on human red blood cells.

The seeds contain an antitumorous saponin prosapogenin, which is active against Walker 256 carcinoma in rats and which on acid hydrolysis gives entagenic acid, D-glucose, L-arabinose, and D-xylose. Acid hydrolysis of the crude saponin liberates methanethiol. Another triterpene sapogenin has been isolated and identified as echinocystic acid. The seeds also contain *O*-(β-D-glucopyranosyl)-L-tyrosine, an unstable aromatic amine, lupeol, and β-sitosterol, while the pericarps contain β-sitosterol, α-amyrin, quercetin, gallic acid, and cyanidin chloride.

Mimosa pudica L.
Family Mimosaceae (Fijian: Cogadrogadro)

Mimosa pudica is a semiprostrate thorny plant that is introduced and naturalized and of widespread distribution. *M. pudica* is often called the "sensitive plant" since the leaves will close if touched.

The leaves with those of other species are used for urinary infection, while the roots are also used for urinary disorders.

An amino acid mimosine, which has been isolated from the plant, is identical with leucaenine from *Leucaena leucocephala*.

Piper methysticum Forst. f.
Family Piperaceae (Fijian: Yaqona)

Piper methysticum is a robust shrub that is commonly cultivated for its lateral roots and thickened underground portions that are powdered and used in the preparation of yaqona, a popular beverage throughout Fiji. The drink, prepared with ceremonial rites by adding the powder to water, is not intoxicating but rather is a mild narcotic, acting as a sedative or soporific. The drink is also used for kidney and bladder troubles, as a diuretic, and often as a panacea for a variety of common complaints. A decoction of the powdered roots is reputed to be used by mothers who have recently given birth to prevent their conceiving again, while the leaves are chewed and swallowed as a contraceptive. The plant is used medicinally in a number of Pacific countries.

The roots contain a series of α-pyrone derivatives, namely kawain, 7,8-dihydrokawain, 5,6-dehydrokawain, yangonin, 5,6,7,8-tetrahydroyangonin, methysticin, 7,8-dihydromethysticin, and eight minor related compounds. A detailed chemical and pharmacological analysis has been reported by Keller and Klohs. The major physiologically active principles in the resin are dihydrokawain and dihydromethysticin, which possess local anesthetic, antifungal, sleep-producing, anticonvulsive, and spasmolytic (smooth muscle contraction) properties. The plant also contains the alkaloid pipermethystine.

Polygala paniculata L.
Family Polygalaceae (Fijian: Ai Roi Ni Turaga)

Polygala paniculata is a common delicate herby weed of American origin. The Fijian name means flag flower, and the flowering spikes are dried for ornaments.

The root is used for toothache and neuralgia.

The root contains methyl salicylate (oil of wintergreen), which explains its use for neuralgia since the compound has marked analgesic effects and antiseptic properties. The roots of the related Chinese medicinal plant *Polygala tenuifolia,* which is used as a sedative and tonic, contains onjisaponin F, a saponin that possesses sedative action.

Bruguiera gymnorhiza (L.)
[syn. *Bruguiera conjugata* (L.) Merrill]
Family Rhizophoraceae (Fijian: Dogo)

Bruguiera gymnorhiza is a very common tree of mangrove swamps that has knobby subsurface roots that stick out from the mud. The fruit germinates while still on the tree, producing an angular cigarlike structure (Figure 5). The wood is hard and durable and makes excellent firewood, and the bark provides a dye.

The bark is used with those of several other species for treating cancer. The bark is also used for syphilis and with the root of vadra *(Pandanus odoratissimus)* for venereal disease.

The bark contains the sulfoxides bruguierol and isobruguierol, which may be the active principles. The plant gives negative tests for alkaloids but the bark is rich in tannins. Recent investigations have shown that tannins are often responsible for the antitumor activity of crude extracts of some plants, especially those having activity against W 256, S 180 and Lewis lung tumors. So far, those tannins examined have not shown sufficient activity to warrant preclinical pharmacology.

Figure 5. Bruguiera gymnorhiza *(L.) (syn.* Bruguiera conjugata *[L.] Merrill) (family Rhizophoraceae) (Fijian: Dogo).*

Atuna racemosa
(syn. *Parinari glaberrima*)
Family Rosaceae (Fijian: Makita)

Atuna racemosa is a common tree that has leathery leaves used for thatching houses and a hard rough fruit that when grated is used to scent coconut.

The fruit is used to treat suppurating sores.

The oil of the fruit when saponified yields a series of aliphatic long chain acids, for example, palmitic, stearic, oleic, linoleic, eleostearic, and parinaric acid (9,11,13,15-octadecatetraenoic acid), and in the nonsaponifiable fraction β-sitosterol, stigmasterol, and 2,3-dihydrobrassicasterol. None of these compounds appear to have healing properties.

Rhizophora samoensis Salvoza (syn. *Rhizophora mangle* L.) Family Rhizophoraceae (Fijian: Tiri)

Rhizophora samoensis is a common many-branched tree with characteristic descending aerial roots that is found in mangrove swamps. The bark is used as a dye.

The bark is reputed to be used medicinally. The root is used for diabetes and for a relapse and with the leaves of other species for loss of appetite and "inner cleansing".

The bark contains tannins and the leaves have antimicrobial activity. The coloring matter of the seeds has also been investigated.

Morinda citrifolia L. Family Rubiaceae (Fijian: Kura)

Morinda citrifolia is a very common small tree, often called the Indian mulberry, that is distributed from India eastward to Polynesia and that is often cultivated for the somewhat tasteless greenish white mature fruits.

The young shoots in coconut oil are greatly esteemed for the cure of ringworm and other similar afflictions such as scabies and the itch and for the treatment of acute rheumatic pains. The leaves are chewed and applied as a poultice for inflammation and rheumatism, and a steam bath made from the leaves is used for stiffness. The leaves are also used for boils and gastric ulcer and with the leaves of other species for hemorrhoids and pregnancy pains. The fruit is also used for ringworm, and the bark is used with the leaves of waro *(Cayratia seemanniana)* for sinusitis. The roots with those of other species are used for malnutrition. The plant has many similar uses in other parts of the Pacific area. A drug made from the roots and trunk is reported to have been marketed for its hypotensive properties.

The bark, heartwood, and roots have all been shown to contain a number of anthraquinones, for example, morindin, monoethoxy-rubiadin, and α-methoxyalizarin, both free and as glycosides. More recently the roots have been investigated in detail and have been shown to contain a variety of anthraquinones, for example, nordamnacanthal, morindone, rubiadin, rubiadin 1-methyl ether, and soranjidiol. These anthraquinones are presumably the active principles in the plant since both antimicrobial activity and other physiological properties have been attributed to these compounds.

Citrus grandis (L.) Osbeck Family Rutaceae (Fijian: Moli Kana)

Citrus grandis is a common apparently indigenous species found in the forest and along riverbanks. The large fleshy segmented fruit (pummelo) is up to 10 cm in diameter and has a pink flesh.

A decoction of the bark is said to be esteemed for its tonic qualities and for its use in the treatment of asthma and as a superior sort of laxative. The bark is also said to have contraceptive and anticancer properties.

The fruit contains the 7-neohesperidosides of naringin, poncirin, neohesperidin, and naringenin 4-glucoside. The plant is also reported to contain the alkaloids putrescine, (–)-stachydrine, quinoline, and tryptamine and tyramine and has recently been shown to contain *N*-methylatanine and preskimmianine. The fruit yields an essential oil that contains limonene as the major component.

Euodia hortensis J.R. et G. Forst. forma *hortensis* Family Rutaceae (Fijian: Uci)

Euodia hortensis is a shrub that is widely distributed in the Pacific Islands. The leaves and flowers are strongly scented and the latter are used to scent coconut oil.

A decoction of the leaves is drunk for difficulties in parturition and for infective hepatitis, while a steam bath of the leaves is used for treating fevers. The bark is also used for treating fevers and is reputed to have contraceptive properties. The plant is also used for a variety of complaints in Tonga and Samoa.

The volatile oil of the leaves contains menthofuran (64%) and evodone; 44 compounds have been isolated by GC–MS from the leaf oil and 23 from the flower oil. The leaves contain the triterpene hortensol and have recently been shown to contain a further ring D secotriterpenoid.

Micromelum minutum Seem.
Family Rutaceae (Fijian: Sasqilu)

Micromelum minutum is a small tree that is common throughout the Fijian Islands. The leaves are chewed as a tonic, and a decoction is used as a superior laxative, for strangulated hernia, or for affected joints. Both the leaves and bark provide an ointment when mixed with coconut oil, which is used to treat carbuncles and skin complaints, respectively.

The bark and leaves contain coumarins; for example, the leaves yield micromelin and microminutum, each of which display in vivo activity in the P-388 lymphocytic leukemia test system. The plant also contains the alkaloid flindersine.

Santalum yasi Seem.
Family Santalaceae (Fijian: Yasi)

Santalum yasi is a small tree known as "Fiji sandalwood" that was once common. The exploration of Fiji is said to have started with the sandalwood trade, which between 1800 and 1805 quickly used up the timber.

The leaves of the plant are used in Samoa for elephantiasis and filariasis, and the wood is used extensively in Fiji and in Polynesia for scenting medicinal oils.

The essential oil of the kernels contains the C_{18} acetylenic acid ximenynic acid; some acetylenes have antibiotic properties.

Pometia pinnata J. R. et G. Forst.
Family Sapindaceae (Fijian: Dawa)

Pometia pinnata is a common tree that is also found in Malaysia, New Guinea, and Polynesia. An extract of the leaves was used in early times as a black hair dye.

The bark has been credited with antifertility properties and is used for diabetes and as a diuretic. An infusion of the bark and leaves together with the leaves of wakai *(Ipomoea batatus)* has been said to be a successful cure for sterility.

The bark contains a saponin that gives oleanolic acid on hydrolysis and also contains leucoanthocyanidins or condensed tannins. The presence of the saponins may explain to some extent the medicinal properties, since some saponins can cause antifertility effects in female mice.

Solanum uporo Dunal
Family Solanaceae (Fijian: Borodina)

Solanum uporo is a bushy shrub that is often cultivated in villages and that is also found in Tonga, Rarotonga, and Tahiti. The leaves make a good pot herb and in earlier days were wrapped around human bodies before cooking.

The leaves are mixed with those of capsicum and rubbed on parts affected by rheumatism. The crushed fruit is used in a similar manner, and they are also eaten for urinary disorders. The juice of the bark is applied to sore eyes.

The plant contains steroidal alkaloids that can be hydrolyzed to afford solasodine. Like diosgenin from *Dioscoria* species, solasodine can be converted into 16-dehydropregnenolone from which all of the major classes of steroid drugs can be produced. Quite apart from their use as raw materials for the steroidal hormone industry, *Solanum* alkaloids may have medicinal value in their own right. For example, in clinical trials solasodine has been shown among other things to improve gastric and hepatic functions with minimal side effects and to show cortisone-like activity in humans.

Dendrocnide harveyi Chew
(syn. *Laportea harveyi* Seem.)
Family Urticaceae (Fijian: Salato)

Dendrocnide harveyi is a large forest tree with leaves that are covered with stinging hairs. Contact with the leaves or sap can cause severe skin irritation that can last for a long time and that is aggravated by contact with water.

The bark is said to provide a good cure for the sting of its own nettlelike leaves. A decoction of the bark is used for treating urinary and menstrual disorders. In the latter case the woman is required to abstain from bathing in seawater and from eating fish. The bark is also used for infected testes and with the barks of other species for arthritis, including rheumatoid arthritis.

The stinging hairs of *Dencrocnide* species contain acetylcholine, histamine, and 5-hydroxytryptamine (serotonin). Acetylcholine is a parasympathomimetic agent for peripheral vascular disease, for intestinal or bladder atomy, for paroxysmal tachycardia, and in quinine, tobacco, or alcohol amblyopia. Acetycholine has also been employed in retinal artery occlusion and occasionally in hypertension. Histamine is used for hyposensitization therapy in allergies, as a counterirritant in ointments, for acute multiple sclerosis, and for other medicinal purposes.

Zingiber zerumbet Sm.
Family Zingiberaceae (Fijian: Cagolaya)

Zingiber zerumbet is an introduced herb that is the only wild ginger in Fiji. *Z. zerumbet*'s scales or bracts turn a rich crimson in the hurricane season and hence the local name "hurricane flower".

The rhizomes are used for colds, indigestion, stomachache, bladder trouble, and blood pressure complaints. The rhizomes are also used for diabetes, to treat fish poisoning, and sometimes for infant's cough. The root is used for a furred tongue and for stomach disorders, and the seeds are often used as a laxative.

The volatile oil from the rhizomes contains a series of terpenes and zerumbone, which is present in greater amount than in the oil from Indian plants. Zerumbone has spasmolytic action and is antistimulating, antibacterial, and antihypertensive (i.e., it lowers the blood pressure and respiration rate); however, in dogs small doses increase the blood pressure. Zerumbone is alleged to have no toxic effect.

Glossary

Abortifacient Anything used to cause an abortion.
Amblyopia Dimness of vision without obvious change in the eye.
Analgesic Pain reliever that does not induce loss of consciousness.
Antimitotic A substance that stops the division of cells.
Antipruritic An anti-itching agent.
Antipyretic An agent that relieves or reduces fever.
Antiseptic Preventing infection or putrefaction.
Artery occlusion Obstruction of an artery.
Bacillary Pertaining to bacilli, any of a large class of straight rod-shaped bacteria.
Bladder atony Abnormal relaxation of the bladder.
Carminative A remedy for flatulence.

Cathartic Producing evacuation of the bowels.

Conjunctivitis Inflammation of the mucous membrane lining of the eyelids.

Consumption A wasting disease, specifically pulmonary tuberculosis.

Diuretic An agent that increases urine flow by acting on the kidneys.

Elephantiasis The final chronic stage of filariasis characterized by a thickening and hardening of the skin and enlargement of the affected part.

Emetic A medicine used to induce vomiting.

Emmenagogue A substance that stimulates or renews menstrual flow.

Epilepsy A chronic nervous condition characterized by sudden loss of consciousness, sometimes accompanied by paroxysmic seizures.

Filariasis A disease caused by nematode worms that affects the lymph gland and connective tissues.

Gastric Pertaining to the stomach.

Hemolytic Pertaining to blood.

Hepatic Pertaining to the liver.

Hypertensive Excessively high blood pressure.

Hypotensive Excessively low blood pressure.

Parasympathomimetic Mimicking the action of that part of the nervous system originating in the cranial and sacral regions of the spinal cord.

Paroxysmal tachycardia Convulsive abnormal rapidity of the heartbeat.

Parturition Childbirth.

Pediculicide An agent acting against pediculosis.

Pediculosis The condition of being infected with lice.

Peptic Of the digestive tract.

Peripheral vascular disease A disease pertaining to surface blood vessels.

Phagocytosis The destruction and absorption of bacteria or microorganisms by phagocytes.

Piscicidal Fish killing.

Purgative An agent causing active evacuation of the bowels that is more drastic than a laxative or aperient.

Reticuloendothelial Netlike layer of cells lining blood vessels.
Scabies Contagious itching skin disease.
Sclerosis The morbid thickening and hardening of a tissue, especially of the arteries.
Soporific Inducing sleep.
Sudorific Producing copious perspiration.

0939-1/86/0089$06.50/0

Hibiscus rosa-sinensis (Reproduced with permission. Copyright 1981 Reference Publications, Inc.)

Chapter 5

Medicinal Plants of Papua New Guinea

DAVID HOLDSWORTH*†

Throughout the world, from the very earliest times of prehistory, men and women have used plants that they found near their villages as medicines. More than 5000 years ago, the ancient Egyptians used sycamore bark and oil from the cedar tree. The Mesopotamians used castor oil, cedar, turpentine, henbane, mint, poppy, fig, and mandrake. More than 1000 medical herbs, including the tranquilizer rauwolfia, were known in India. The Chinese wrote manuscripts about their medicines and mentioned rhubarb, castor oil, camphor, and cannabis.

European medical doctors used many of these medicines, and the exploration of new lands about 500 years ago increased the number of plant medicines known. Several important ones came from the Americas, including the purgative ipecacuanha, the important antimalarial quinine, and the narcotic tobacco.

A rich heritage of traditional knowledge of the uses of plants as medicines can be found in Papua New Guinea. Since man first arrived probably about 60,000 years ago, he used the plants in his immediate area. Some were found useful for building houses, canoes, and weapons. Others were found to be edible and were

*Chemistry Department
University of Papua New Guinea
Papua New Guinea
†Current address: Chemical Education Sector. University of East Anglia, Norwich, England

used as foods. Systematically, by trial and error, some plants were found to be poisonous and others were discovered to be effective in curing diseases or body ailments.

The useful medicinal plants were remembered, and the information was told to successive generations. In some areas one man in the village would act as a professional healer. He would teach his son his accumulated knowledge, and the position of village healer would be inherited. Often the medicine man would be a sorcerer, and the use of magical phrases, psychology, impressive ritual, and fear would go hand in hand with the use of medicinal plants and poisons.

The island of New Guinea is the second largest in the world. Papua New Guinea, the eastern half of New Guinea with the large islands New Britain, New Ireland, Manus, and Bougainville, has a population of three million with more than 700 languages and cultures. Each community, until recently isolated by rugged terrain and hostility, has developed its own medical practice using plants immediately available.

Papua New Guinea is now experiencing rapid changes. Young men and women leave their villages for school and work. Many traditional customs are disappearing or are being replaced. The effectiveness of a shot of penicillin or other modern medicines at the rural health center or hospital has meant that the role of plant medicines is diminishing and traditional medical practice is, in many areas, quickly disappearing. Older villagers often complain that their sons and daughters are not interested in learning ancestral medicinal secrets, so they do not continue the custom of passing on the accumulated knowledge of the clan. Recording the traditional uses of medicinal plants throughout the whole of Papua New Guinea is essential.

IMPORTANCE OF MEDICINAL PLANTS

In the last 100 years, many important discoveries were made in the newly developing science of chemistry. Medicinal plants were examined carefully, and many substances were extracted from different parts of plants. A medicinal plant might contain one or many different compounds that might have medical activity. These pure compounds could be used or mixed together to make very effective medicines. The correct dose prescribed by a doctor, nurse, or health worker would ensure effective medical treatment. More recently, the chemical structures of the pure extracted compounds have been found. The manufacture of these compounds or similar compounds is possible; this fact allows medicine to become cheaper and available to most people throughout the whole world.

However, some traditional medicines may cause unwelcome side effects. For instance, the bark of the willow tree was used in Europe for hundreds of years to treat headaches. Unfortunately, the bark gave some patients diarrhea. Chemists extracted and purified a compound from the bark, and its structure was found. A research worker discovered that if a small chemical change was made, the new compound could cure headaches faster with no side effects. Also, this compound was cheap and easy to manufacture. It is known as aspirin and is now used throughout the world.

Plants used as medicines in Papua New Guinea might contain substances that might lead to new advances in medicine, and new drugs might become available to help not only the people of Papua New Guinea but also the world.

RECORDING THE USES OF MEDICINAL PLANTS IN PAPUA NEW GUINEA

For more than 10 years the University of Papua New Guinea has carried out a survey of the plants used as medicines in many different provinces of the country where there has been local interest. Collections have been made by science graduates or students who have first discussed the use of plants with their relatives or village elders. Later, the village was visited to collect specimens of the plants and to discuss and record their uses.

In some instances professional traditional headers cooperated in the survey. These men usually insisted that the plants should not be seen by other people in the village.

To date, more than 1000 plants have been collected. Most have been identified by the staff of the Papua New Guinea Herbarium in Lae. Nearly 400 different species of plants have been recorded for a wide variety of ailments.

For each plant, the local village name, details of the part of the plant used (leaf, root, bark, etc.), how the plant is used, and the dosage and the frequency of dosage are noted. All plants are tested in the Chemistry Department, University of Papua New Guinea, for the presence of alkaloids by extraction and then with precipitating reagents or thin-layer chromatography with suitable locating sprays. Perhaps later, many of the plants will be tested by pharmacologists to monitor biological activity and possible toxic effects. Then new drugs might be introduced from plant material, and chemical analysis may lead to synthetic medicines.

Records indicate that plants have been used to treat malaria, coughs, asthma, body pains, sores, tropical ulcers, burns, wounds, insect bites, diarrhea, dysentery, stomachache, constipation, poisoning, fatigue, eye infections, ear infections, sore throats, headaches, venereal diseases, grille, swollen spleen, mouth ulcers, sore noses, rashes, scabies, snakebites, toothache, vomiting, intestinal worms, and gout. Plants are also used by women in childbirth and as contraceptives. In several instances if the plant grows in other southeast Asian countries or Pacific islands, its usage is identical or very similar.

Some examples of the medicinal uses of plants are given.

Alstonia scholaris

Alstonia scholaris is a tree with white, strongly perfumed flowers. An infusion of the leaves is used to treat a fever or a headache at Hisiu near Yule Island. An infusion of the dried bark is stated to be more powerful and is used to treat a severe fever or malaria. The bark is used internally for fevers in Morobe, stomachache in the Sepik, dysentery and diarrhea on Manus Island, and abdominal and chest pains in the Northern Province. A drink from the sap of the stem when squeezed into water is used to treat a cough on Normanby Island and gonorrhea in Milne Bay. The barks of *A. scholaris, Cananga odorate,* and *Endospermium labios* are scraped together, heated over a fire, and squeezed, and the juice is drunk to treat gout in the North Solomons. Waste scrapings are bandaged onto the swelling overnight. *A. scholaris* is also used to treat malaria in the same area.

The sap is used to ease constipation in Buang villages. At the Plitty village hospital, a decoction of the dried bark is used extensively to treat asthma, hypertension, pneumonia, and lung cancer. The dried bark is used extensively to cure fevers in Indonesia, Malaysia, and the Philippines *(1)*. A tincture or infusion of the dried bark is listed as a treatment of malaria in the *Extra Pharmacopoeia (2)*. Leaves were used to cure fevers in Irian Jaya *(3)*, and Webb *(4)* reported the use of *A. scholaris* to treat fever, dysentery, and abdominal pains in New Ireland. In the Philippines a decoction of the bark is used to treat fevers *(5)*. The bark is used to combat diarrhea in Indonesia *(6)*.

Barringtonia asiatica

The fresh nut of *Barringtonia asiatica* is scraped and applied directly to a sore. The dried nut is ground into a powder, mixed with water, and drunk to treat coughs, influenza, sore throat, bronchitis, diarrhea, and a swollen spleen after malaria. *B. asiatica* is used on wounds and to treat tuberculosis in Samoa *(7)*. Scraped barks of *B. asiatica* and an unidentified tree (kakaii) are heated in a bamboo container with water. The concentrated juice is tasted daily, and betel nut is chewed to treat tuberculosis. The medicine, now used near Kieta, was introduced from Malaita, Solomon Islands.

Calophyllum inophyllum

In Manus, leaves of *Calophyllum inophyllum* are heated over a fire until soft and applied to ulcers, boils, cuts, and sores. On Dobu Island, leaves are boiled and a skin rash is washed periodically with the solution. At Kerigia leaf sap is mixed with water and drunk to relieve dysentery. The tree finds uses in Fiji for skin inflammations *(8)*, in New Caledonia for ulcers, wounds, and sores *(9)*, and in Samoa for skin infections and scabies *(7)*.

Cassia alata

For treatment of the skin fungus grille *(Tinea imbricata)*, leaves of *Cassia alata* are squeezed until soft and rubbed regularly onto the affected part of the body. The use of the plant is widespread throughout Papua New Guinea. The plant was recorded by Bailey *(10)* to be used to cure eczema. A decoction of the leaves, bark, and yellow flowers was used to bathe the affected parts. In Surinam, the crushed leaf is used for skin diseases, and in Tanzania, leaves are crushed in water to bathe children's skin eruptions *(11)*. The leaves are used in Malaysia and Indochina to treat ringworm *(1)*. Leaf extracts have been shown by Atkinson *(12, 13)* to give positive antibiotic tests.

Coix lachryma-jobi

A patient with suspected cancer or other internal sores or internal parasites is given a drink of the juice from heated young stems mixed with traditional salt (extracted from wood ashes). At Kieta, a concoction drunk by women from crushed leaves and water is used to induce fertility after each period.

Crassociphalum crepidiodes

Sores and new cuts are moistened by the juices squeezed from heated young leaves. The treatment may be repeated, if necessary, in 2 or 3 days. In Africa, the powdered leaf is used to stop bleeding *(11)*. The leaf is crushed and applied to a sore in Milne Bay.

Desmodium

Crushed leaves of *Desmodium umbellatum* and shoots are used to massage an enlarged spleen caused by malaria. A decoction of the leaves is used in Motu villages as a general sickness preventative. The solution is used to bathe the body to prevent a slight chill from developing into a fever. Crushed leaves and stem are applied directly onto burns and topical ulcers. In the Moluccas, the plant is considered to be an astringent and is used by women after childbirth *(1)*. In the Moli District, Guadalcanal (Solomon Islands), for a child with a hard abdomen, young leaves are heated over a fire and extracted with water. Small portions of this medicine are given to the child to drink three times daily *(14)*. Two species of *Desmodium* are used for fevers in Guam *(15)*. A decoction of *Desmodium styracifolium* is used by the Chinese to clear a fever and to treat chronic hepatitis *(16)*.

Euodia bonwickii

A patient with malarial fever drinks a decoction of the leaves and soft stalk twice daily until cured. Young leaves heated over a fire in a wild ginger leaf and squeezed may be mixed with lime and applied to the forehead to give relief to a headache or slight fever. Various species of *Euodia* are used to treat fevers in Samoa *(7)*, Malaysia *(1)*, and India *(17)*.

Fagraea

Fagraea bodenii leaves chewed with traditional salt are considered to cure an enlarged spleen caused by malaria. Warriors chew leaves as a stimulant before battle. In Malaysia, *Fagraea racemosa* root and leaves are used as a tonic after fever. Heated leaves may also be applied to the abdomen to treat malaria and other fevers *(1)*. The bark of *Fagraea imperialis* and the root bark of *F. racemosa* are taken by fever patients in India *(17)*.

Galbulimima belgraveana

Galbulimima belgraveana bark is chewed and spat into a bamboo container and salt is added. The spittle is swallowed to bring relief to abdominal and other body pains. Powdered bark mixed with wild tobacco *(Nicotiana tabacum)* and ginger *(Zingiber officinale)* is rubbed into the head to treat hair lice. Cribb *(18)* states that the bark is chewed by warriors before tribal fights to produce violent intoxication and hallucinations followed by extreme drowsiness.

Hibiscus

The sap from the squeezed leaves of *Hibiscus schizopetalus* is drunk to induce labor. Flowers and young leaves of *Hibiscus rosa-sinenis* are soaked in coconut water, and the solution is given to induce labor in the Northern Province and Samoa *(7)*.

Impatiens

In Kabiufa, the whole plant of *Impatiens hawkerii* is cooked and eaten by children with stomachache. The juice from the fruit and the leaves is rubbed onto the legs of small children who are retarded in their walking. The Kukukuku people near Marawaka mix the leaves of *I. hawkeri* with those of *Coleus scutellarioides* to rub on the stomach of a pregnant woman to give some relief from labor pains. At Aseki, a pregnant woman may chew young leaves and salt to induce labor. The dried stem of *Impatiens balsalmina* is used by women with labor difficulties in China *(19)*.

The Sukima tribe in East Africa uses *Impatiens walleriana* as an abortifacient *(11)*.

Lunasia amara

Young leaves are chewed with betel nut and lime. A weekly dose is stated to act as a contraceptive. A daily dose is said to induce sterility. Regular washing with a solution obtained by squeezing old leaves in water kills fleas on a human or a dog. In Indonesia, a decoction of the bark and leaves is used externally for swollen limbs and against skin diseases *(6)*. The leaves and bark are strongly alkaloidal, and over 10 alkaloids have been extracted and identified *(20)*. Uphof *(21)* reports that the bark contains two alkaloids that have a paralyzing effect upon the heart.

Melia

Melia dubia leaves are boiled in water, and the solution is used to treat many internal sicknesses, for instance, diarrhea. *Melia azerdach* is used for abdominal pains in China *(16)*, in New Caledonia *(9)*, and in Guam *(15)*.

Polyscias

The top leaves of *Polyscias scutellaria* are boiled and given to a sickly child as a tonic. In New Caledonia, *Polyscias pinnata* leaves are used to treat exhaustion *(22)*. A mixture of root and leaf of *P. scutellaria* is used as a diuretic in Guam *(15)*.

Scolopia novoguineansis

The whole plant is boiled, and the solution is drunk to treat a severe cough. The treatment is repeated if the cough persists. The leaves are also used as a contraceptive. A woman who has recently given birth to a child will eat no meat or fish for 5 days. She then makes a payment to a traditional healer who gives her a concentrated solution of the leaves of *S. novoguineansis* and a ginger root. One drink is stated to give temporary infertility.

Synedrella nodiflora

In Central Province coastal villages, disjointed bones are massaged daily with sap from squeezed leaves. For treatment of diarrhea on Buka, some of the root is chewed with betel nut and lime and then swallowed. The rest of the root is left to smoke in the kitchen over a fire. In Indonesia, leaves are used to treat stomachache *(1)*.

Tournefortia

The leaf of *Tournefortia sarmentosa* is chewed by a patient suffering from malaria. The leaf is also used in a cure for stomachache. In Sepik River villages, leaves are eaten with food to give relief to stomachache. *Tournefortia montana* is used to treat fever in Vietnam *(1)*.

Vitex nugundo

A cough or a sore throat is treated by drinking the mixture formed when the sap of the leaves is squeezed into water. A decoction of the leaves is drunk to give relief to catarrh in India *(17)*. Blackwood *(23)* reported the root was used as an oral contraceptive by women of the northern Bougainville island.

CHEMICAL PROPERTIES OF MEDICINAL PLANTS IN PAPUA NEW GUINEA

About 2% of all plants randomly tested were found to contain alkaloids. However, 10% of plants used in traditional medicine and over 35% of the plants used internally to combat malaria were alkaloid positive *(24)*.

An alkaloid virosecurinine was isolated from the leaves (and roots) of *Securinega methanthesoides* (syn. *Securinega virosa)*, a decoction of which has a bitter taste and is used in a Central Province village to cure fever. The Nyamwezi and Shambala tribes of East Africa drink an infusion of the roots with meat broth to treat malaria *(11)*. Purified virosecurinine was shown by optical rotation measurements as well as infrared, nuclear magnetic, and mass spectroscopy to be a diastereoisomer of allosecurinine, $C_{13}H_{15}NO_2$ *(25)*. Virosecurinine was shown to be mildly toxic to mice with an LD_{50} of 73 mg/kg of body weight. Death resulted from violent tonic convulsions and paralysis similar to those observed with strychnine poisoning. Virosecurinine had no effect on the growth of the bacteria *Escherichia coli, Staphylococcus aureus,* and *Bacillus subtilis* or the fungi *Fusarium moniliforme, Penicillium viridicatum, Arachniotus niger, Rhizoctonia solani, Rhizopus stolonifer,* and *Curvularia lunata aeria.* Its low solubility in water probably is the reason that poisoning has not been reported among patients *(26)*.

Severe burns are treated in the Gazelle Peninsula by application of the sticky sap of *Aloe camperii,* believed to be introduced in the last 100 years from the Horn of Africa. *Aloe aborescens* has been reported to be used for burns in Palau. Recently, *Aloe vera,* used for stomach ulcers and burns in the Western Highlands of Papua New Guinea, has been proven effective in healing X-ray burns *(19)*. *A. vera* contains the *C*-glycosylchromone aloesin and its aglycon aloesone, both of which absorb ultraviolet light strongly and are probably responsible for the healing properties *(27)*. Aloin, also present in *A. vera,* is a strong laxative. Probably *C*-glycosylaloin is hydrolyzed and oxidized in the stomach to aloe emodin, which has a powerful purgative effect but can cause vomiting when taken orally.

A medicinal plant garden will soon be established at the Wau Ecology Institute to enable larger quantities of plant material to be available for pharmacological testing. Overseas universities and institutes are being contacted to see how they might cooperate with screening processes, pharmacological testing, and chemical analysis.

It is hoped that rural health centers will have a medicinal plant garden nearby so village people can choose between traditional and Western medicine. Should Western medicine be temporarily out of stock, plant material would be always available as a substitute. The emphasis of the survey has been to collect plants for identification and information on their uses before this knowledge is lost. This information will thus be available in 20 or 100 years for research when the people in an area may have forgotten some or all of their ancestral knowledge. Future Papua New Guineans may then be able to direct medical research from traditional knowledge that would otherwise have been forgotten.

LITERATURE CITED

1. Burkill, I. H. "A Dictionary of the Economic Products of the Malay Peninsula"; Ministry of Agriculture: Kuala Lumpur, 1966; Vols. 1-2.
2. Martindale, W. "Extra Pharmacopoeia," 25th ed.; Pharmaceutical: London, 1967; p. 1502.
3. Warburg, G. "Neu-Guinea"; Krieger, M., Ed.; Schall, Berlin, 1899; pp. 36-72.
4. Webb, L. J. *Proc. R. Soc. Queensl.* **1960,** *71(16),* 103-10.
5. de Padua, S. L.; Lugod, G. V.; Pancho, J. V. "Handbook on Philippine Medicinal Plants"; Univ. of the Philippines: Los Banos, 1977; Vol. 1.
6. van Steenis-Kruseman, M. J. "Select Indonesian Medicinal Plants," Bulletin No. 18; Organization for Scientific Research in Indonesia: Djakarta, 1953.
7. Uhe, G. *Econ. Bot.* **1974,** *28(1),* 1-30.
8. Zepernick, B. "Arnzneipflanzen der Polynesier"; Dietrich Reimer Verlag: Berlin, 1972.
9. Rageau, J. "Travaux et Documents de l'ORSTOM"; Office de la Recherche Scientifique et Technique Outre-Mer: Paris, 1973.
10. Bailey, F. M. "Comprehensive Catalogue of Queensland Plants"; Government Printer: Brisbane, 1909.
11. Watt, J. M.; Breyer-Brandwijk, M. G. "The Medicinal and Poisonous Plants of Southern and Eastern Africa"; Livingstone: London, 1962.
12. Atkinson, N. *Aust. J. Exp. Biol. Med. Sci.* **1946,** *24,* 49.
13. Atkinson, N., *Aust. J. Exp. Biol. Med. Sci.* **1956,** *34,* 17-26.
14. Foye, F., unpublished manuscript, Honolulu, 1976.
15. Haddock, R. L. South Pac. Comm. Reg. Tech. Meet. Med. Plants, Working Paper 15, Papeete, Tahiti, 1973.
16. "Chinese Paramedical Manual," translation; Running: Philadelphia, 1977.
17. Chopra, R. N.; Chopra, I. C.; Nayar, S. L. "Glossary of Indian Medicinal Plants"; C. S. I. R.: New Delhi, India, 1956.
18. Cribb, A. B. "Wild Medicine in Australia"; Collins: Sydney, Australia, 1981.
19. Perry, L. M. "Medicinal Plants of East and Southeast Asia; M. I. T. Press: Cambridge, Mass., 1980.
20. Boit, H. G. "Ergebnisse der Alkaloid-Chemie bis 1960"; Akad-Verlag: Berlin, 1961.
21. Uphof, J. C. T. "Dictionary of Economic Plants"; von Cramer: Lehre, 1968.
22. Sterly, J. "Heipflanzen der Einwohner Melanesiens"; Arbeitsstelle fur Ethnomedizine: Hamburg, 1970.
23. Blackwood, B. "Both Sides of the Buka Passage"; Charendor: Oxford, 1935.
24. Holdsworth, D. K. "Medicinal Plants of Papua New Guinea," Tech. Paper No. 175, South Pacific Commission, Noumea.
25. Holdsworth, D. K. *Proc. Papua New Guinea Sci. Soc.* **1970,** *22,* 41-45.
26. Hill, L.; Holdsworth, D. K.; Small, O. R. *Papua New Guinea Med. J.* **1975,** *18(3),* 157-61.
27. Holdsworth, D. K. *Planta Med.* **1971,** *19(4),* 322-25.

0939-1/86/0101$06.00/0

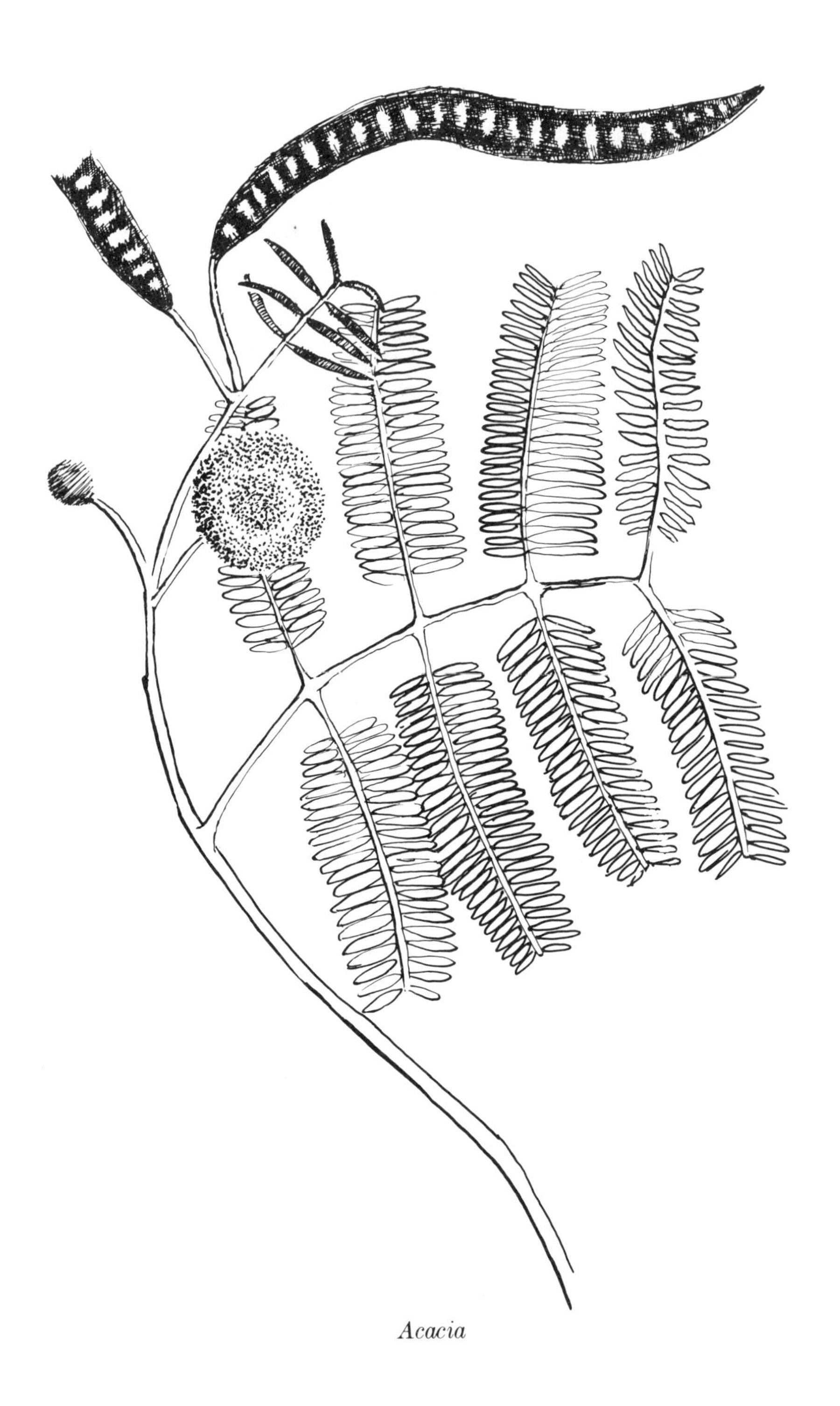

Acacia

Chapter 6

Australian Medicinal Plants

GEOFFREY N. VAUGHAN*

Australian medicinal plants can be subdivided into three major categories: (1) plants used by the Aborigines, (2) plants used by European settlers, and (3) native Australian plants that have been used medicinally and yield pharmacologically active compounds. Other classifications could be used. For example, many introduced species and imported plant materials are used today as herbal medicine, but this chapter will concentrate mainly on native plants.

Australia is a vast continent with a unique flora that has evolved to survive in the wide physical and climatic conditions across the country. The *Acacia* and *Eucalyptus* species grow in physical conditions ranging from desert to snow and from rain forest to temperate conditions. *Acacias* can be found as small shrubs only centimeters tall or as large trees 15 m high. The uniqueness and the variety are certainly outstanding characteristics of Australian plants.

PLANTS USED BY THE ABORIGINES

The Australian Aborigine settled the continent some 40,000 years ago from a migration path that came out of Asia. The land mass separated to create a new group of people who had to develop

*Victorian College of Pharmacy
381 Royal Parade
Parkville, Victoria 3052 Australia

their own methods for the treatment of disease using the available resources.

Medical treatment evolved and depended on spiritual beliefs and the development of herbal medicine. In the development of people across the globe, individual cultures had always used the spiritual and physical approaches to medicine side by side even though their cultures had not crossed for centuries and no form of communication existed between them.

The Australian Aboriginal system depended on a tribal base. Tribes would develop their own customs and cultures and would be extremely independent and in many ways extremely individual. Yet a common thread ran through the tribes, as it had done with the different races across the world, and that thread was a common approach to medicine using divine power and plant products as two distinct approaches to therapy.

The spiritual approach was left to the medicine man. He would be an elder in the tribe who had been selected through breeding and tribal evaluation as the man who could bring about miracles. Through his distinctive attire, his chanting, and his dancing, he could drive away evil spirits from the body and restore normal health to a sick member of the tribe.

Herbal Treatments

At the herbal level, medicines were developed by trial and error and passed on by word of mouth from generation to generation. Herbal treatment did not have to work miracles because the Aborigines only experienced simple disorders. These disorders included headaches, digestive complaints, antiseptic needs, ophthalmia, fever, rheumatic pain, skin eruptions, respiratory complaints, and dysentery. Treatment of sprains, wounds, and burns was also needed. The life-threatening diseases of humans were not experienced by the early settlers of Australia.

No history of hypertension or heart attack can be found. Contagious and infectious diseases such as smallpox, cholera, malaria, rabies, tuberculosis, and venereal disease were not known to the Aborigine. They did not even experience the childhood diseases of measles and mumps.

European settlement, which began in 1788, brought disease with it, causing havoc in the Aboriginal population which had not built up any natural immunity through previous exposure.

Medical Plant Groups

Historians, pharmacists, and doctors have been challenged to piece together the plant medicines of the Aboriginal people. The major difficulties have been the following: (1) The tribal pattern of living, the vastness of the continent, and the native plants varying from region to region meant that a great variety of plants were used to treat any given disease. Sometimes one plant might be used to treat different diseases by different tribes. (2) The lack of written records and the large number of different dialects, through which the medical treatments were passed on from generation to generation, have caused difficulties in piecing together the true picture. (3) Different "dosage from design" was used by different tribes. One would use the leaves, another the bark; one would use extracts, another would use powdered plant material; one would use internal medicine, another topical application. The possible combinations of these factors have led to further difficulties in understanding medical customs.

Although vast gaps exist in our knowledge, much work has been done to record the plants of medical importance. Table I lists a selection of plants that have been used in a selection of treatments. The list is not comprehensive and is presented to give an indication of the variety of plant species that have been used.

Table I. Plants in Aboriginal Medicine

Plant	*Disorder*
Acacia (wattle)	common ailments
Ajuga (Australian bugle)	skin ulcers, gangrene
Alphitonia (white ash)	body pain
Amyema (maiden's mistletoe)	inflammation
Centipeda (sneezeweed)	sore throats, eyes
Cleome (tickweed)	intestinal worms
Dioscoria (native yam)	skin cancer
Eucalyptus (gum tree)	wounds, coughs, colds, sprains, rheumatism, fever, headache, sore eyes, diarrhea
Euodia (toothache tree)	toothache
Ficus (native fig)	wounds, infection, ringworm
Melaleuca (paperbark)	colds, general sickness
Mentha (river mint)	coughs, colds, tonic spasms
Nymphaea (waterlily)	tonic spasms, skin disorders
Pittosporum (pittosporum)	colds, internal medicine
Sterculia (peanut tree)	sore eyes
Urtica (nettle)	rheumatism, sprains

Plants for Pleasure

Narcotics and euphorics have been used by humans all over the world for thousands of years. A comparatively small number of plants have evolved as the major source of narcotics. The Aborigines were no exception in seeking pleasure from plants, although the varieties of plants used were quite numerous, which again reflects the separation of tribes, the size of the continent, and the variations in the botanical regions. A comparatively comprehensive list of those plants now known to have been used as narcotics is given at the bottom of this page.

By far the most important species in the list of narcotics is *Duboisia hopwoodii* because of its reasonably large geographic distribution and its high activity. The active principles are nicotine and nornicotine, which are present at 1–2% levels (compared to tobacco, which has a higher content of 4–6%).

The Aborigines, like all other races, sought a love potion. Legend indicates that many species were tried, but evaluation shows that none of the species had any special properties other than cosmetic and perfume effects. Some of the plants that have been used as aphrodisiacs are listed on page 107.

Aboriginal Medicine: Narcotics

Amorphophallus (stinking arum)
Callicarpa longifolia (chuckin)
Dodonaea (wild hops)
Duboisia hopwoodii (pituri)
Duboisia myoporoides (corkwood)
Eucalyptus (gum tree)
Galbulimima belgraveana (argara)
Isotoma petraea (rock bluebell)
Nicotiana (native tobacco)
Papaver somiferum (opium poppy)
Amanita muscaria (fly agaric)
Copelandia cyanescens (blue meanies)
Psilicybe cubensis (gold top)

Aboriginal Medicine: Aphrodisiacs

Arbarema grandiflora (fairy paintbrush)
Banalophora fungosa (drumsticks)
Lycopodium phlegmaria (tassel fern)
Phallus rubicundus (stinkhorn)
Pittosporum venulosum
Viscum articulatum (leafless mistletoe)

PLANTS USED BY EUROPEAN SETTLERS

The east coast of the Australian continent was first discovered by the British navigator Captain James Cook in 1770. He reported his finding and the existence of the Aborigines on his return to Britain. The British government decided to settle the continent through the export of convicts. The first convict fleet arrived in 1788 under the command of Captain Arthur Phillip, thus starting the European settlement and the gradual rise of European customs. The earliest settlers were military personnel, convicts, and some administrative and professional support people. Doctors arrived with the first fleet and brought with them European medicine (some of which had already developed from trade with India, the Dutch East Indies, and other Asian trading ports). As Aboriginal customs became known to the white settlers, they began to use some Aboriginal medicines. Sometimes this usage was of necessity because of the uncertainty of supplies arriving on infrequent ships.

In the early 19th century, "free" settlers arrived to take the benefit of a new life in a new continent that promised growth and wealth. Gold was discovered in the 1840s, and this finding caused a population surge that has never really ceased even to the present day. Migrants to Australia have been predominantly European; thus, the European tradition has been firmly established over almost two centuries of influence. Early settlers brought with them prepared medicines or seeds and plants of medicinal herbs to ensure sources of supply.

Through this relatively short period of time, modern medicine evolved and rapidly replaced most of the herbal treatments and diminished the interest in any development of Australian plant medicines by the population at large. Also the Aboriginal influence was declining rapidly because of a lower population and the "takeover" of the continent by the new settlers.

Available records indicate that the Europeans used plant medicine from three sources: (1) Aboriginal medicine (*Acacia, Eucalyptus,* etc.), (2) traditional medicine from Europe (*Capsella, Chicorium, Mentha,* etc.), (3) plants from Europe and elsewhere that had not been used in herbal medicine but that had been tried on an experimental basis and had gained some popularity (*Nasturtium, Plantago, Ranunculus*). Today most medicines used in herbal treatment are imported to Australia from many overseas sources and are part of the increasing popularity of alternative medical treatment.

NATIVE AUSTRALIAN PLANTS USED IN MODERN MEDICINE

Although plant preparations and crude extracts are not widely used in Australia, interest in the Australian flora as sources for biologically active material and drug substances is increasing. The more important plants and their active ingredients are shown in Table II.

Some of the plants listed in Table II are extracted and purified in Australia and processed into pharmaceutical products (e.g., morphine alkaloids) for the Australian and export markets. Plants are also exported in dried form for processing overseas to obtain the pure drug substance (e.g., hyoscine).

Another relationship exists whereby overseas countries have established plantations of Australian plants in various countries and through growing, processing, extraction, and purification the countries use the active ingredient in their home markets. The interesting example in this category is *Solanum aviculare* (kangaroo apple) as a source of the steroidal alkaloid solasodine in Russia. Solasodine can be easily converted chemically to pregnenolone as a source of synthetic sex hormones and corticosteroids. Australia has never developed the kangaroo apple industry because of the availability of pregnenolone derivatives in Western countries from the Mexican yam.

Table II. Plants in Modern Medicine

Plant	*Active Ingredient*
Bauerella simplicifolia (scrub yellowwood)	acronycine
Bursaria spinosa (blackthorn)	aesculin
Calvatia gigantea (giant puffball)	calvacin
Cantharanthus roseus (periwinkle)	vincristine
Cryptocarya glabella (poison walnut)	cryptopleurine
Datura stramonium (thorn apple)	hyoscyamine
Digenea simplex (wormweed)	kainic acid
Digitalis purpurea (foxglove)	digitoxin
Duboisia myoporoides (duboisia)	hyoscine
Eucalyptus macrorhynca (red strinybark)	rutin
Eucalyptus species	eucalyptus oil
Gracilaria verrucosa (sea noodles)	agar
Macrocystis pyrifera (giant kelp)	alginate
Persoonia (geebung)	antibacterial
Phormidium	growth-stimulating substance
Papaver somniferum (opium poppy)	morphine
Silybum marianum (variegated thistle)	tyramine
Solanum aviculare (kangaroo apple)	solasodine
Solanum hermanii (devil's apple)	alkaloid
Stephania joponica (tape vine)	aknadine
Tylophora credbriflora (coast tylophora)	tylocerbrine

CONCLUSION

This chapter reviews Australian medicinal plants and their role in Aboriginal medicine, the medical interface between the Aborigines and European settlers, and the more recent developments in economic terms. In summary, Australian medicinal plants have (1) had a rich history that developed with the tribal structure of Aboriginal life, (2) served a purpose in early European settlement, and (3) produced some interesting drugs and related compounds. These pharmaceutical products have not been highly developed by Australia because of its population size and have contributed to the economy mainly as part of the total export industry.

GENERAL REFERENCES

1. Cribb, A. B.; Cribb, J. W.; "Wild Medicine"; Fontana: Sydney, 1981.
2. Lassak, E. V.; McCarthy, T. "Australian Medicinal Plants"; Methuen: Sydney, 1983.

0939-1/86/0109$06.00/0

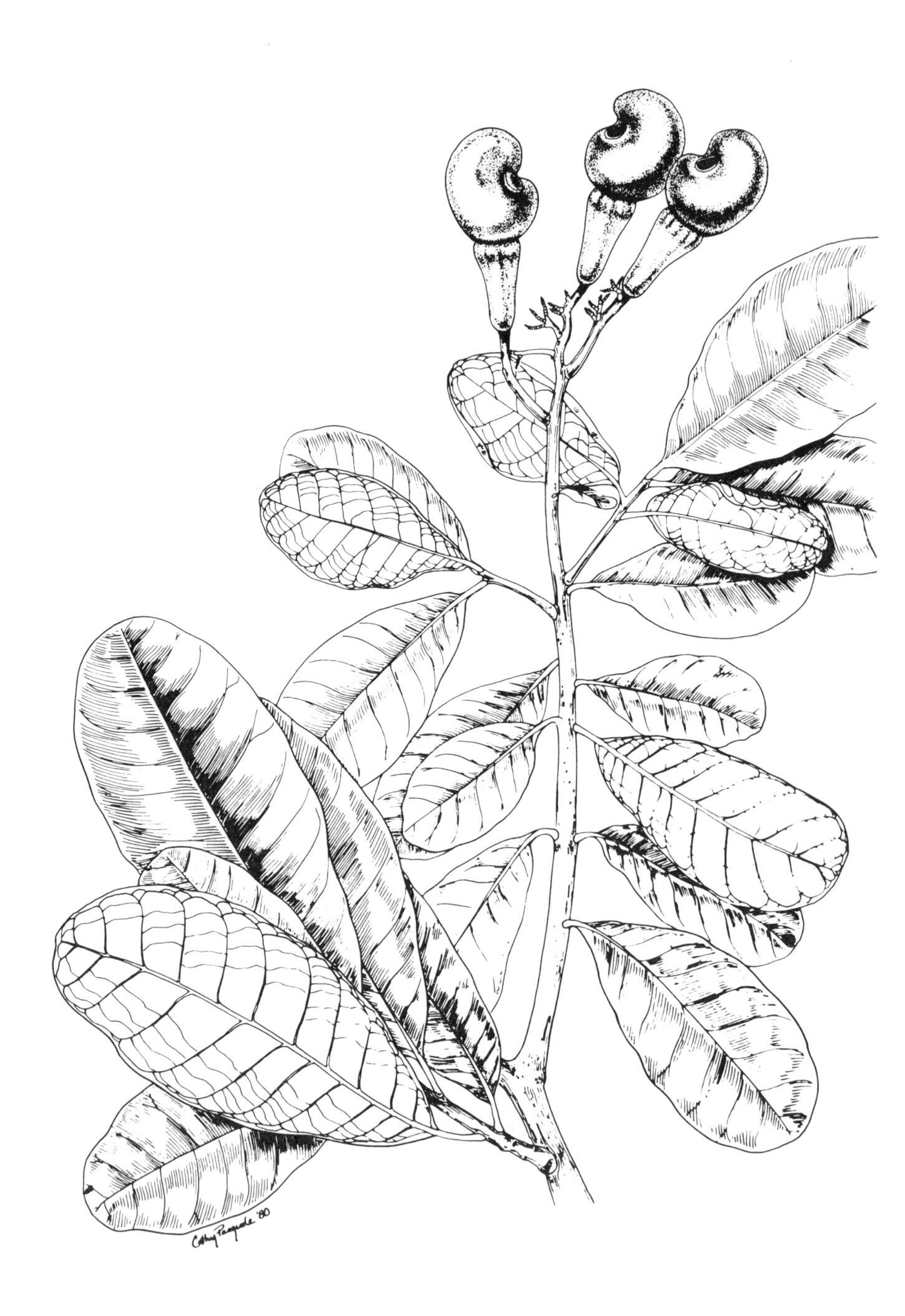

Anacardium occidentale (Reproduced with permission. Copyright 1981 Reference Publications, Inc.)

Chapter 7

Plants Used in African Traditional Medicine

KURT HOSTETTMANN*
and ANDREW MARSTON*

The continent of Africa provides a rich source of plants that, over the centuries, have found use in curing all manner of ailments. In the last 100 years or so, a large number of their medicinal uses have been documented, and literature for the areas of southern Africa *(1)*, eastern Africa *(2)*, and western Africa *(3)*, to give just a few representative examples, is available. However, regions remain in which traditional medicine has been somewhat neglected, and the great danger is that with the gradual decline in the importance of traditional healers, valuable information on medicinal plants will be lost. This problem is aggravated by the fact that Western medicine is often considered superior to local remedies, and the direct use of plants is pushed into the background. To counteract this trend, certain African countries have introduced measures to preserve the traditional healer network so that prescriptions that have been passed down from generation to generation will continue to be employed. The central African state of Malaŵi is such a case in point.

In Malaŵi, an attempt to promote traditional medicine alongside modern medicine has been made, and this attempt has

*School of Pharmacy
University of Lausanne
1005 Lausanne, Switzerland

resulted in the formation of the Herbalists' Association of Malaŵi (HAM), an organization that claims to have more than 50,000 members *(4)*. HAM operates closely with the Malaŵian Ministry of Health and with the University of Malaŵi, meaning that information is freely available on the use of plants in a variety of different ailments. In collaboration with the Department of Chemistry of the University of Malaŵi, our institute is currently carrying out an investigation on various plant species employed by members of HAM, in an effort to characterize the compounds responsible for the therapeutic effects claimed by the traditional healers. The organization of the program is such that the healer supplies a specimen of the plant he uses for a particular local remedy, and this specimen is subsequently identified by a qualified botanist. Once the species of plant is known, isolation of active principles is carried out. The activities claimed by the healer are frequently confirmed after pharmacological testing, and in this way useful leads into new drugs are possible. This approach is less time consuming (and less expensive) than random screening of plant material.

The acceptance of traditional healers in the modern health system of Malaŵi means that the confidence of these men and women is considerably easy to obtain. They are less suspicious of officials investigating their remedies and are much more willing to provide exact details of the plants they use.

Despite the positive aspects of this cooperation, some problems occur. Diagnoses, especially for internal complaints, often lack precision, and what may, for example, be considered as a cancer of the stomach may, in fact, be an ulcer. As Sofowora points out *(5)*, the traditional healer tends to treat the symptoms rather than the disease. Healing may in certain cases be a result of the placebo effect rather than the therapeutic activity of a preparation. One must also be careful of some rather exaggerated claims of traditional medicine, and a good case in point is that of the fruits of *Kigelia africana* (Bignoniaceae), claimed in eastern Africa to aid enlargement of the male genitals. The shape of the fruit explains the use of the common name "sausage tree", and in this instance, what is referred to is probably something akin to the doctrine of signatures rather than a true medical effect. The other local uses of *K. africana*—a purgative and dressing for ulcers—are much more relevant to the pharmacological investigation of this plant.

The limitations of traditional medical practice must not, however, detract from the enormous benefits that it has provided. Probably the best example to illustrate this point is the centuries-old use in mental illness of *Rauwolfia serpentina* (Apocynaceae) in India. Many alkaloids have now been isolated from Indian and African *Rauwolfia* species, and reserpine, obtained from the roots, is of vital importance for its sedative and antihypertensive properties.

The list of biologically active principles isolated from African medicinal plants is growing all the time, and a selection of activities particularly interesting to our group will be described in this chapter.

MOLLUSCICIDAL ACTIVITIES

Molluscicidal or snail-killing activities of plants are of special importance in the widespread tropical disease known as schistosomiasis *(6)*. Schistosomiasis or bilharzia affects millions of people living in African, Asian, and South American countries and is closely linked with certain species of aquatic snails. These snails play host to miracidia, which in turn hatch from eggs desposited by humans suffering from the disease (Figure 1). One way of attacking the diease is to eradicate the host snails, and the other two viable alternatives are either to use orally administered drugs to kill schistosomes in infected persons or to employ a combination of chemotherapy and molluscicide. This latter approach is probably the most effective, but a number of plants (or plant-derived compounds) have proved very effective in snail-control schemes *(7)*. Some of these phytochemical compounds are saponins, and the history of their use as molluscicides arises from interesting observations that saponin-containing plants have been used since ancient times as soap substitutes or fish poisons (Table I).

Crude aqueous extracts of *Phytolacca dodecandra* (Phytolaccaceae) berries provide some of the most active molluscicides; their toxicity to the snails approaches that of commercially available synthetic products *(6)*. The way in which the activity of the berries was first noticed came from their use as a soap substitute in Ethiopia. The incidence of snails was dramatically reduced where people washed their clothes, and after much

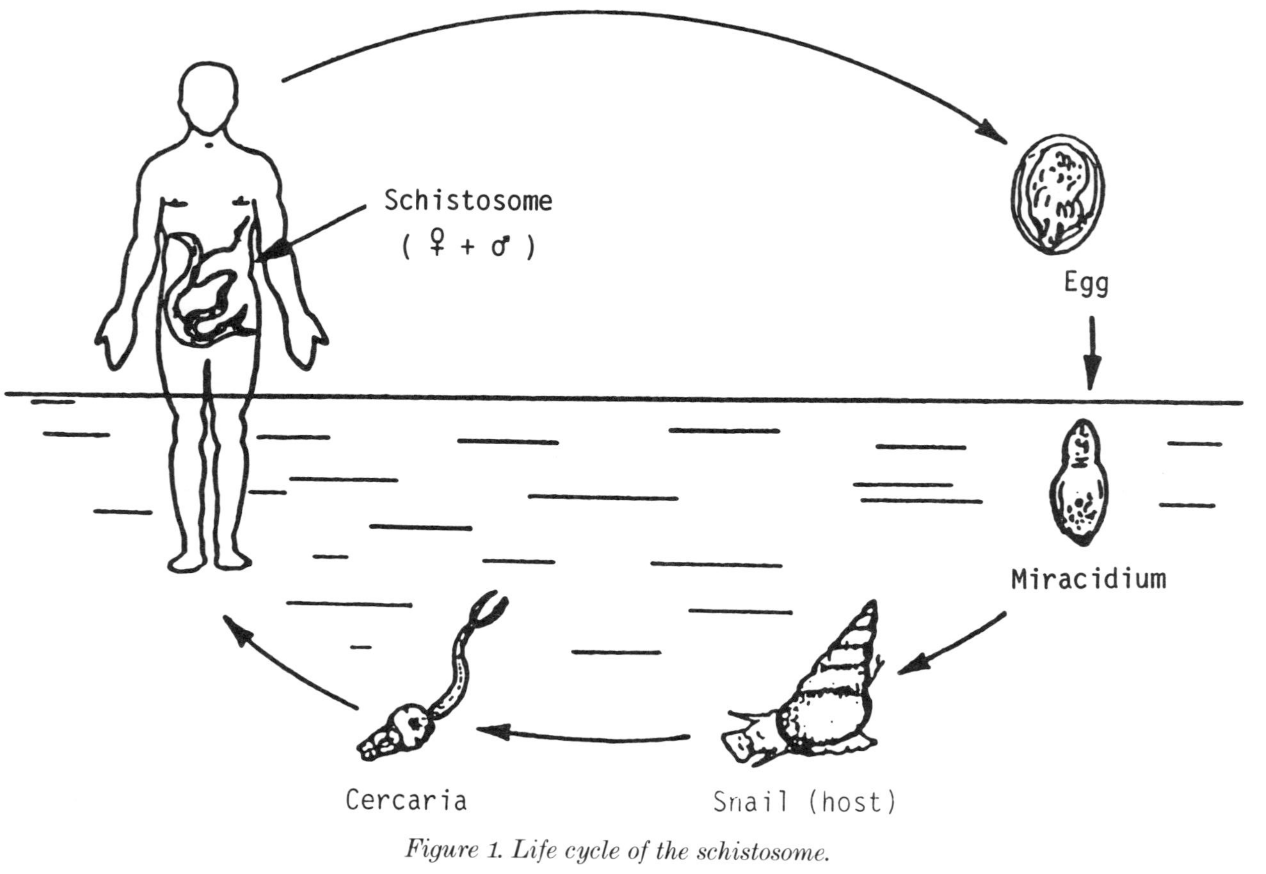

Figure 1. Life cycle of the schistosome.

Table I. Some Saponin-Containing Plants Used as Fish Poisons or Soaps

Plant	*Fish Poison*	*Soap Substitute*
Phytolacca dodecandra (Phytolaccaceae)	—	X
Swartzia madagascariensis (Leguminosae)	X	—
Sapindus saponaria (Sapindaceae)	X	X
Securidaca longepedunculata (Polygalaceae)	X	X
Sesbania sesban (Leguminosae)	X	—
Balanites aegyptiaca (Balanitaceae)	X	—
Neorautanenia pseudopachyrhiza (Leguminosae)	X	—
Securinega virosa (Euphorbiaceae)	X	—
Strychnos innocua (Loganiaceae)	X	X

investigation, the monodesmosidic triterpene saponins of *P. dodecandra* were shown to be responsible for these effects *(8, 9)*. These saponins are glycosides of the aglycons oleanolic acid, hederagenin, and bayogenin (Table II).

Another saponin-containing plant *(Talinum tenuissimum,* Portulacaceae), used in Malaŵi traditional medicine as a schistosomiasis cure, was found to be very active against snails of the genus *Biomphalaria.* A total kill of snails in 24 h was achieved at concentrations of 25 μg/mL for extracts of the tuber *(10)*. On fractionation of these extracts, the saponin responsible for the activity (1.5 μg/mL after 24 h) was isolated.

Swartzia madagascariensis is a tree that belongs to the family Leguminosae and enjoys a wide reputation as a medicinal plant in southern Africa *(1)*; it has uses that vary from emetic to insecticide to abortifacient. In addition, the long pods of the tree are a very effective fish poison and are used as such over large areas of Africa *(1)*. Following Mozley's suggestion that *S. madagascariensis* should be used as a vegetable molluscicide *(11)*, the chemistry of the molluscicidal components from the pods has been under investigation in our laboratory, and a series of structurally fascinating saponins have been isolated. The activities of these saponins are such that field trials, in conjunction with the Swiss Tropical Institute, Basel, have already been carried out in a bilharzia-infested region of Tanzania. The trials have shown that a water extract of the ripe pods is molluscicidal at very low concentrations. Factors such as ecological impact, biodegradability,

Table II. Molluscicidal Saponins of *Phytolacca dodecandra*

Compound	R^1	R^2	R^3	*Moluscicidal Act. at 24 h (μg/mL)*
1	H	Rha-^{2}Glc-^{2}Glc-	H	3
2	H	Glc-^{4}Glc- $\vert^{2}$ Glc	H	3
3	H	Glc-^{4}Glc- $\vert^{3}$ Gal	H	1.5
4	H	Glc-^{4}Glc- $\vert^{2}$ Glc	HO	12
5	HO	Glc-^{4}Glc	HO	12

and distribution have all been studied, with the result that *S. madagascariensis* has now been shown to have real potential as a solution to problems of schistosomiasis in focal areas. The tree is found abundantly in infected areas, often in close proximity to transmission sites. Consequently, transport problems do not arise, and because the pods are easy to pick and exist in large quantities, application problems are minimized.

The molluscicidal activity of naphthoquinones has recently been reported for the first time from our laboratory *(12)*. Extracts of root bark from *Diospyros usambarensis,* a bush of the Ebenaceae family found in central Malaŵi, were found to kill schistosomiasis-transmitting snails. On fractionation of the ligroin extract, a large amount of 7-methyljuglone was isolated, and this naphthoquinone was found to be molluscicidal at 5 μg/mL. Following the discovery of this activity, a number of other naphthoquinones were investigated and found to have similar molluscicidal potential (Table III). Dimeric naphthoquinones isolated from *D. usambarensis* were inactive *(12)*.

The introduction of a hydroxyl group into the quinonoid ring caused a significant diminution in the molluscicidal potential. Menadione, a synthetic prothrombogenic drug, was also tested and, interestingly enough, gave a very high activity.

Several species of the genera *Lonchocarpus* (Leguminosae) and *Tephrosia* (Leguminosae) have been traditionally used as fish poisons, and *Tephrosia vogelii*, widespread in central Africa, is well known for its piscicidal and insecticidal properties, due to the presence of rotenoids. Preliminary tests showed extracts of the leaves of *T. vogelii* to have molluscicidal activity, but the rotenoids subsequently isolated, deguelin and tephrosin, were inactive *(13)*. These rotenoids are virtually insoluble in water, and test solutions are invariably in the form of suspensions.

Other plant-derived natural products can also be toxic to mollusks *(6)*, and the most potent molluscicides are to be found among the alkenyl phenols *(14)*. An alkenyl phenol with a triply unsaturated C_{15} side chain, from cashew nut shells *(Anacardium occidentale,* Anacardiaceae), is active against snails at 0.7 μg/mL and would seem to provide the ideal naturally occurring molluscicide, if the dermatitis-inducing effects in humans were not so pronounced. The activity has proved high enough all the same to justify the introduction of field trials in Mozambique with the shell oil—a byproduct of cashew nut production. Tannins, often found in large quantities in numerous species of plants, are a potential source of molluscicides. A number of common tannin-

Table III. Molluscicidal Activities of Selected Naphthoquinones

Compound	*Molluscicidal Act.*[a] *(μg/mL)*
7-Methyljuglone	5
Plumbagin	2
Juglone	10
Isojugloine (lawsone)	50
3-Methoxy-7-methyljuglone	>50
Vitamin K_3 (menadione)	3
Naphthazarin	50

[a]One hundred percent snail kill after 24 h.

containing plants have been investigated in our group *(15)*, and although their efficacy is high, a great deal of testing remains to be done.

SCHISTOSOMICIDAL ACTIVITIES

A number of plants are used in Malaŵi traditional medicine to cure patients who pass blood in their urine. As this phenomenon is often associated with people suffering from the urinary form of schistosomiasis, this phenomenon can be used to indicate plants that may be used to kill schistosomes in the body. Included in these remedies are the roots of *Glycine javanica* (Leguminosae) and *Vigna radiata* (Leguminosae), which are boiled in water; the decoction is then drunk twice a day to effect a cure.

In addition to these plants, a large number of additional plants are used by herbalists in the cure of bilharzia itself. The extent of the disease in Malaŵi can be appreciated when one considers the number of peparations that exist. We have tested a number of extracts of these plants for their schistosomicidal activity, and positive results were obtained in vitro for *Diospyros usambarensis* (Ebenaceae), *Phytolacca dodecandra* (Phytolaccaceae), and *Mammea africana* (Guttiferae). In the hope that the molluscicidal naphthoquinones from *D. usambarensis* might also prove to be schistosomicidal, we tested 7-methyljuglone in vitro, but the results were rather disappointing. However, vitamin K_3 gave some activity in vivo, and investigation of further naphthoquinones might be worthwhile. Obvious applications can be found for compounds that are both molluscicidal and schistosomicidal because in a bilharzia control scheme, the disease could be attacked at two focuses by the *same* product and the chances of preventing transmission would therefore be enhanced.

FUNGICIDAL ACTIVITIES

Specially selected hybrids of cereal crops and other plant food sources often suffer from a decreased natural resistance to environmental attack by insects, viruses, and fungi. Therefore, to avoid outbreaks of disease that could eliminate a crop, spraying with fungicides and pesticides is necessary. Common human diseases caused by fungi, for example, *Trichoderma,* need also be controlled. The genetic versatility of microorganisms and corresponding problems of resistance mean that new antimicrobial and antifungal substances are continually being sought.

Among the types of organisms pathogenic to plants are fungi, bacteria, viruses, mycoplasms, and nematodes. These organisms cause a whole range of diseases, which also include mildews and rusts. In our laboratory, we have chosen as a preliminary screen for fungicidal activity the spores of *Cladosporium cucumerinum,* which is a fungus pathogenic to cucumbers and yet harmless to humans—harmless, therefore, to the personnel carrying out the tests. Our bioassay is a modification of the method of Homans and Fuchs *(16)* and involves the migration of a test substance with a suitable solvent over a normal aluminum-backed silica gel thin-layer chromatographic (TLC) plate. After removal of excess solvent from the TLC plate, a suspension of *C. cucumerinum* spores in a medium containing maltose and salts is sprayed over the plate and the whole is incubated in the dark at 25°C for 3 days under humid conditions. Examination of the plate after this time interval shows it to be covered with a gray growth of spores, except in those places to which the fungicidal compounds have migrated; here the white silica gel of the plate shows. A modification of this TLC bioassay allows minimum inhibitory concentration (MIC) values to be measured *(17)*. In this case, wells are filled with a mixture containing spores, medium, and substance under test; MIC values are taken from those wells containing the lowest concentration of fungicide required to stop growth of *C. cucumerinum.*

The twigs of *Diospyros usambarensis* (Ebenaceae) are used in Malaŵi as chewing sticks to keep teeth clean and to avoid mouth infections. Consequently, this plant was selected as a potential source of antimicrobial and antifungal compounds. Indeed, testing ligroin and chloroform extracts of the root bark in the *C. cucumerinum* bioassay proved the presence of very active antifungal compounds. Isolation of these substances showed them to be naphthoquinones (except shinanolone), the major constituent being 7-methyljuglone, the most active of the compounds tested *(17)* (Table IV).

Various other species of the genus *Diospyros* from Malaŵi, for example, *D. zombense, D. whyteana,* and *D. kirkii*, are very active against *C. cucumerinum* and are presently under investigation in

Table IV. Antifungal Activities of *Diospyros usambarensis* Naphthoquinones in the *Cladosporium cucumerinum* Bioassay

Naphthoquinone	*Minimum Act. on TLC Plate (μg)*	*MIC (μg/mL)*
7-Methyljuglone	0.025	2
2-Methoxy-7-methyljuglone	0.5	20
3-Methoxy-7-methyljuglone	0.3	20
Isodiospyrin	0.125	5
Mamegakinone	10	n.t.[a]
Shinanolone	5	n.t.

[a]n.t. = not tested.

our laboratory. Additional Malaŵi medicinal plants have given positive results in the antifungal test and are undergoing further examination.

Warburganal is an antifungal principle from the leaves of *Warburgia stuhlmannii* (Canellaceae) and *Warburgia ugandensis,* trees used widely in east African traditional medicine *(18)*. Not only is warburganal antifungal but also it shows wide-spectrum biological activity, being an antifeedant, antibiotic, antiyeast, molluscicidal, and hemolytic compound.

ANTITUMOR ACTIVITIES

An amazing amount of effort has been put into the search for antitumor agents from higher plants, lower plants, marine organisms, and animals *(19)*. The National Cancer Institute (NCI) approach in the United States was one of random screening until the program was recently drastically modified, despite the discovery of several tumor inhibitors that are already in clinical trials. Furthermore, a whole variety of different structural types that showed some sort of antitumor or cytotoxic activity were isolated. Another approach is to derive hints of anticancer activity from plants used in traditional medicine *(20)*. The chances of finding worthwhile results by this method appear to be greater than by random selection, and this fact suggests a correlation between folklore and those plants with positive anticancer activity *(20)*. Lists of plants having some recorded use in the treatment of various forms of malignant growth have been published (for example, Ref. *21*); these provide useful sources of information for phytochemists wishing to isolate the active principles. However, the diagnosis of a "cancer" is not always foolproof, and disorders such as internal ulcers, swellings, or abscesses are often quite falsely described as tumorous growths. Conversely, what is diagnosed as an abscess may actually prove to be a tumor, and the plant responsible for its treatment will naturally be applicable to the study of possible antitumor constituents. An example of a false diagnosis is frequently encountered when traditional healers attempt to cure bleeding in women between menstrual periods. This bleeding is quite falsely attributed to "uterine cancer," and the therapy is therefore not of an antitumor nature.

Numerous plants used in African traditional medicine have been investigated for their cytotoxic and antineoplastic properties. Three examples will be described here, to give just a small insight into the possibilities that African plants have for supplying new leads.

The first example concerns *Psorospermum febrifugum* (Guttiferae), a bush found over wide areas of central and eastern Africa. As its name implies, this plant has fever-reducing properties, the investigation of which is presently being undertaken in our laboratory. Additional uses of this plant include leprosy treatment and the cure of skin diseases *(1)*. Guided by an in vitro bioassay, Kupchan et al. *(22)* isolated an antitumor xanthone, psorospermin, from the roots. Psorospermin was only present at a level of 0.004% but was cytotoxic to 9KB cells at $ED_{50} = 0.1\ \mu g/mL$ and showed activity in the in vivo P-388 mouse leukemia model at $T/C = 140–160\%$ (doses of 0.1–8 mg/kg) *(22)*. In addition, the xanthone was active in mammary and colon solid tumor models *(23)*. Further xanthones have since been isolated from *P. febrifugum.* Two are inactive in the P-388 in vivo test, while another has borderline activity.

Other phytochemical investigations *(24, 25)* have shown the roots to contain further substances of interest: anthrones, anthraquinones, and vismiones. These compounds generally have prenylated side chains and are highly colored. One of the compounds, an anthrone, has been shown to possess in vivo activity against P-388 leukemia in mice *(25)*.

We are currently reinvestigating the very interesting antitumor activity of *P. febrifugum* in collaboration with the Swiss

Institute for Experimental Cancer Research and at the same time are studying other members of the Guttiferae found in Africa. As an in vitro test to follow activities during fractionation of plant extracts, we are measuring cytotoxicities to two human colon cancer cell lines. The method employs a chromogenic substrate for a lysosomal enzyme present in the cells, and cell numbers can easily be calculated by spectrophotometric means. Active compounds isolated with the aid of this bioassay can then be tested on xenografts in athymic mice.

Maprounea africana (Euphorbiaceae) is used in eastern Africa as a purgative and cure for syphilis. Alcoholic extracts of the dried roots showed activity against P-388 leukemia in mice *(26)*. Further investigation by the same research group led to the isolation of a number of pentacyclic triterpenes, one of which was highly active in the P-388 in vivo test ($T/C = 163\%$ at 0.4 mg/kg). An in vitro 9PS assay gave ED_{50} values of 2×10^{-4} μg/mL, while there was virtually no activity in the 9KB assay ($ED_{50} > 10$ μg/mL). Structure–activity relationships have shown that the 7β-hydroxyl function is essential for cytotoxicity *(26)*.

One of the most useful drugs in the chemotherapy of acute childhood leukemia is a plant-derived product: the alkaloid vincristine from *Catharanthus roseus* (Apocynaceae). This plant, found in Madagascar, is reputed among the locals to be effective against diabetes, and during investigation of its pharmacological effects, the antileukemic activity was discovered *(27)*.

CONCLUSIONS

A brief insight has been given into the different (and important) biological activities of African medicinal plants. Hundreds of these plants have been tested in diverse biological systems, and if further vital active principles, such as the alkaloids from *Catharanthus roseus* (Apocynaceae), are to be discovered, this testing must be continued and, at the same time, diversified. Useful indications from traditional medicine considerably aid the search for biological activities, but the right pharmacological test is not always easy to find and refinements are often required. However, the correlation between the activity claimed by local healers and the activity revealed by pharmacological screening systems is high. Following leads from Malaŵi herbalists, we have been able to isolate compounds with molluscicidal, schistosomicidal, and fungicidal activities from the relevant plant sources. To broaden the scope of the program, we are currently extending biological testing to analgesic, antipyretic, and antiinflammatory activities, and the indications are that here too pharmacological data corroborate the precisions of the traditional healer. Therefore, when the diagnoses provided by traditional healers are correct, a great proportion of their corresponding plant-based cures are reliable. This finding implies that the plants involved represent an invaluable source of lead compounds waiting to be developed, for use not just as pharmaceuticals but also as local preparations for distribution in African countries themselves.

LITERATURE CITED

1. Watt, J. M.; Breyer-Brandwijk, M. G. "The Medicinal and Poisonous Plants of Southern and Eastern Africa"; Livingstone: Edinburgh, 1962.
2. Kokwaro, J. O. "Medicinal Plants in East Africa"; East African Literature Bureau: Nairobi, 1976.
3. Dalziel, J. M. "The Useful Plants of West Tropical Africa"; The Crown Agents for the Colonies: London, 1956.
4. Msonthi, J. D. *J. Ethnopharmacol.* in press.
5. Sofowora, A. "Medicinal Plants and Traditional Medicine in Africa"; Wiley: Chichester, 1982.
6. Marston, A.; Hostettmann, K. *Phytochemistry,* **1985,** *24,* 639.
7. Kloos, H.; McCullough, F. S. *Planta Med.* **1982,** *46,* 195.
8. Parkhurst, R. M.; Thomas, D. W.; Skinner, W. A.; Cary, L. W. *Phytochemistry* **1973,** *12,* 1437.
9. Domon, B.; Hostettmann, K. *Helv. Chim. Acta* **1984,** *67,* 1310.
10. Gafner, F.; Msonthi, J. D.; Hostettmann, K. *Helv. Chim. Acta,* **1985,** *68,* 555.
11. Mozley, A. "Molluscicides"; Lewis; London, 1952.
12. Marston, A.; Msonthi, J. D.; Hostettmann, K. *Planta Med.* **1984,** *50,* 279.
13. Marston, A.; Msonthi, J. D.; Hostettmann, K. *Phytochemistry* **1984,** *23,* 1824.
14. Sullivan, J. T.; Richards, C. S.; Lloyd, H. A.; Krishna, G. *Planta Med.* **1982,** *44,* 175.
15. Schaufelberger, D.; Hostettmann, K. *Planta Med.* **1983,** *48,* 105.
16. Homans, A. L.; Fuchs, A. *J. Chromatogr.* **1970,** *51,* 325.
17. Marston, A.; Msonthi, J. D.; Hostettmann, K. *Farm. Tijdschr. Belg.* **1984,** *61,* 291.
18. Nakanishi, K. In "Natural Products and Drug Development"; Krogsgaard-Larsen, P.; Brøgger Christensen, S.; Kofod, H., Eds.; Munksgaard: Copenhagen, 1984; p. 31.
19. Schepartz, S. A. *Cancer Treat. Rep.* **1976,** *60,* 975.
20. Farnsworth, N. R.; Kaas, C. J. *J. Ethnopharmacol.* **1981,** *3,* 85.
21. Hartwell, J. L. "Plants Used Against Cancer"; Quarterman: Lawrence, Mass., 1982.

22. Kupchan, S. M.; Streelman, D. R.; Sneden, A. T. *J. Nat. Prod.* **1980,** *43,* 296.
23. Cassady, J. M.; Chang, C.-J.; Habilo, A. M.; Ho, D.; Amonkar, A.; Masuda, S. In "Natural Products and Drug Development"; Krogsgaard-Larsen, F.; Brøgger Christensen, S.; Koford, H., Eds.; Munksgaard: Copenhagen, 1984; p. 228.
24. Botta, B.; Delle Monache, F.; Delle Monache, G.; Marini Bettolo, G. B.; Oguakwa, J. U. *Phytochemistry* **1983,** *22,* 539.
25. Amonkar, A.; Chang, C.-J.; Cassady, J. M. *Experientia* **1981,** *37,* 1138.
26. Wani, M. C.; Schaumberg, J. P.; Taylor, H. L.; Thompson, J. B.; Wall, M. E. *J. Nat. Prod.* **1983,** *40,* 537.
27. Carter, S. K. *Cancer Treat. Rep.* **1976,** *60,* 1141.

0939-1/86/0124$06.00/0

Chapter 8

Antithrombotic Agent of Garlic: A Lesson from 5000 Years of Folk Medicine

ERIC BLOCK*

As soon as humans had reached the stage of reasoning, they discovered through trial and error plants that might be used as foods; as medicines to alleviate or cure ills; as narcotics in religious rituals; or as poisons for killing enemies, for hunting, or for administration of justice. Folk medicine, consisting largely of the use of herbs and other plants, originated in this fashion and still persists. Among the remedies developed were febrifuges from the bark of the willow tree (genus *Salix*), now known to contain derivatives of salicylic acid (which is related to aspirin), drugs from the Chinese plant rauwolfia for treatment of high blood pressure (the alkaloid reserpine being the active principle), the Chinese herb mahuang *(Ephedra vulgaris)* for treatment of asthma (ephedrine is the active alkaloid component), plasters of moldy bread for sores (penicillin?), quinine from cinchona bark for treatment of malaria, and the heart drug digitalis from the purple foxglove.

In their search for plants of medicinal value, primitive humans naturally investigated those vegetables already in common use

*Department of Chemistry
State University of New York at Albany
Albany, NY 12222

such as garlic. Garlic was well known to ancient and medieval civilizations and was widely appreciated *(1–6)*. Egyptian pharaohs had clay or wooden carvings of garlic placed in their tombs to ensure that afterlife meals would be well seasoned. The Bible mentions that the Jews, wandering for 40 years in the wilderness of Sinai, fondly "remember the fish which we did eat in Egypt so freely, and the pumpkins and melons, and the leeks, onions, and garlic." The pilgrim in *Canterbury Tales* "well loved he garlic, onions, aye and leeks, and drinking of strong wine as red as blood ...." On the other hand, abstinence from garlic is urged by Bottom, in Shakespeare's *A Midsummer Night's Dream,* who tells his troupe of actors to "eat no onions nor garlic, for we are to utter sweet breath," and in *Measure for Measure,* Lucio slanders the Duke by saying that "he would mouth with a beggar, though she smelt brown bread and garlic." The ancient Greeks considered the odor of garlic vulgar and prohibited garlic eaters from worshipping at the Temple of Cybele. Even today notices may be seen at the entrance to Buddhist temples prohibiting on the premises wine as well as garlic, onions, leeks, scallions, and ginger, all of which are considered profane.

Garlic is a member of the lily family (Liliaceae) and goes by the botanical name *Allium sativum* ("allium" may derive from the Celtic word "all," which means pungent). Garlic is among the oldest cultivated plants in existence. Garlic's precise origin, most likely in central Asia, predates written history.

Garlic has been used in folk medicine for thousands of years *(1–6)*. In ancient times when no alternative was available but to seek remedies for illness from natural sources, anything so pungent and strong was considered to have great power. In the *Codex Ebers,* an Egyptian medical papyrus dating to about 1550 B.C., of the the more than 800 therapeutic formulas, 22 mention garlic for a variety of ailments including heart problems, headaches, bites, worms, and tumors. The Phoenicians and Vikings took garlic with them on long voyages as a valuable remedy for the various ailments common to sailors. Dioscorides, chief physician to the Roman Army in the second century A.D., prescribed garlic as a vermifuge, or expeller of intestinal worms. During the first Olympic games in Greece, the athletes were said to have employed garlic as a stimulant (athletes have apparently not changed much through the years!). Garlic has been widely used in Indian medicine in the form of an antiseptic lotion for washing wounds and ulcers, and in this century garlic was used apparently with some success during the world wars as an antiseptic in the prevention of gangrene. Folk medicine is often intertwined with legend, as in the case of "Four Thieves Vinegar". In 1721, the story goes, four condemned criminals were recruited to bury the dead during a terrible plague in Marseilles. The gravediggers seemed to be immune from the plague. Their secret turned out to be a concoction they drank consisting of macerated garlic in wine, which immediately became famous as *vinaigre des quatre voleurs.* The drink is still available today in France. Such tales undoubtedly inspired verses such as those of Sir John Harington in the *Englishman's Doctor:*

> Garlic then have power to save from death
> Bear with it though it maketh unsavory breath,
> And scorn not garlic like some that think
> It only maketh men wink and drink and stink.

Garlic was recommended for its medicinal effects by Aristotle, Hippocrates, and Aristophanes. Numerous therapeutic uses for garlic are listed by the Roman naturalist Pliny the Elder in *Historia Naturalis (1–6)*:

> [Garlic]...keeps off serpents and scorpions by its smell... the ancient used also to give it raw to madmen...it relieves hoarseness if taken thus...when cooked in honey and vinegar it expels tape-worms and other parasites of the intestines...its drawbacks are that it dulls the sight, causes flatulence, injures the stomach when taken too freely, and creates thirst....

Pasteur *(7)* mentioned medicinal and antibacterial properties of garlic in 1858, while Schweitzer *(8)* used garlic in the treatment of amoebic dysentery, cholera, and typhus in Africa. More recent work has shown that garlic juice inhibits the growth of bacteria of genera *Staphylococcus, Streptococcus, Bacillus* (including *B. dysenteriae* and *B. enteritidis*), and *Vibrio* (including *V. cholerae*) at dilutions of 1:125,000 and shows a broad spectrum of activity against zoopathogenic fungi and many strains of yeast *(Candida albicans)* including some causing vaginitis. In the People's Republic of China, 11 cases of cryptococcal meningitis were reportedly successfully treated with garlic extracts *(9)*. Garlic extracts destroy mosquito larvae. Japanese chemists have discovered that garlic increases the capacity of the body to assimilate vitamin B_1, a discovery that provided the basis for development of a useful vitamin pill that contains allithiamin, a chemical combination of a component of garlic and vitamin B_1.

In France horses suffering from blood clots in the legs were fed onions and garlic. More recent clinical studies suggest that garlic is a vasodilator (opening up blood vessels) and that both garlic and onions are anticoagulants. In Switzerland in 1948 Piotrowski demonstrated that in 40 of 100 patients with abnormally high blood pressure, garlic caused a reduction of 20 mmHg or more after about 1 week of treatment *(10)*. Garlic has been claimed to exert a hypoglycemic effect (lowering the level of glucose in blood), which in turn is attributed to a garlic-induced increase in the serum insulin level *(11)*. In India in 1979 Sainani and co-workers *(12)* reported an intriguing epidemiological study of three populations with different dietary habits. The subjects were vegetarians in the Jain community in India who consumed onions and garlic in liberal amounts (at least 50 g of garlic and 600 g of onions per week), in smaller amounts (no more than 10 g of garlic and 200 g of onions per week), or never in their lives. Members of the three groups were matched for sex, weight, age, and social status and had daily diets that were similar in regard to caloric, fat, and carbohydrate intake, with a difference primarily in consumption or abstinence from onions and garlic. The group with total abstinence from onion and garlic had significantly shorter blood coagulation time, poorer fibrinolytic activity, and higher plasma fibrinogen when compared to the other two groups. The mean fasting cholesterol levels for the first, second, and third groups were 159, 172, and 208 mg %, respectively; total triglycerides for the three groups were 52, 75, and 109 mg %; the group consuming liberal amounts of onions and garlic had the lowest mean levels in both cases. The conclusions drawn were that the difference in blood coagulation parameters and cholesterol and triglyceride levels was associated with the ingestion of onion and garlic and that these vegetables exerted a beneficial effect.

Studies during the 1970s and 1980s by several different research groups *(13–17)* suggest that essential oils extracted from onion and garlic inhibit in vivo and in vitro platelet aggregation;

this effect is not seen with most proprietary garlic capsules. Taken together with the studies *(18, 19)* on the prevention by onions and garlic of hyperlipidemia and hypercholesterolemia (high blood lipid and cholesterol, respectively) and the lowering by garlic of low and very low density lipoprotein levels, which are high-risk factors in heart disease, the folklore on the beneficial effects of these alliaceous plants in the prevention of coronary thrombosis, stroke, and atherosclerosis is given some credence.

Chemists have long been attracted to garlic and other members of the allium family because of their strong sulfur odors, sharp taste, and unique physiological effects. Recent research demonstrates that cutting these simple aromatic vegetables releases a number of low molecular weight organic molecules that share the characteristics of possessing the element sulfur in bonding forms rarely found in nature, of showing high reactivity leading to spontaneous conversion to other organic sulfur compounds that undergo still further transformations, and of displaying biological properties as diverse as lacrimatory, antibacterial, antifungal, and anticoagulant effects.

One of the earliest studies of the chemical constituents of garlic was performed in 1844 by the German chemist Wertheim *(20, 21)*, who observed:

> Garlic is well known to be a prized luxury of common man in many lands; namely it is cultivated and consumed in great quantities in Bohemia, Poland and Hungary. Its appeal can be attributed mainly to the presence of a sulfur-containing, liquid body, the so-called garlic oil. All that is known about the material is limited to some meager facts about the pure product, which is obtained by steam distillation of bulbs of *Allium sativum.* Since sulfur bonding has been little investigated so far, a study of this material promises to supply useful results for science.

Ten years prior to Wertheim's work, German chemists Liebig and Zeise *(22, 23)* had reported the first preparation of an organic compound of sulfur, ethanethiol, C_2H_5SH, a liquid with an offensive skunklike odor detectable at concentrations as low as 1 part in 50 billion parts of air. Wertheim in 1844 isolated some strong-smelling volatile constituents from steam-distilled garlic oil (the so-called essential oil of garlic). Wertheim proposed the name allyl (from *Allium*) for the hydrocarbon group in the oil and schwefelallyl ("allyl sulfur" in English) for the volatiles. "Allyl" is still used today to refer to groups of structure $CH_2{=}CHCH_2$. In 1892 another German chemist, Semmler *(24)*, subjected garlic cloves to steam distillation, obtaining 1–2 g of an evil-smelling oil/kg of garlic. This oil was separated by fractional distillation into a number of pure products, which were identified by elemental analysis and chemical transformations. The major component of garlic oil was reported to be diallyl disulfide, $CH_2{=}CHCH_2SSCH_2$-$CH{=}CH_2$, with lesser amounts of diallyl trisulfide and tetrasulfide.

The next key discovery in the study of the chemistry of garlic was made in 1944 by Cavallito and co-workers of the Sterling-Winthrop Company in Rensselaer, N.Y. *(25, 26)*. These researchers noted that a freshly prepared aqueous extract of ground garlic cloves possessed high antibacterial activity when tested by a procedure similar to that used for penicillin. The distilled oil of garlic as studied by Semmler showed no antibacterial activity. Cavallito extracted 4 kg of ground garlic cloves at room temperature with 95% ethanol and, after filtration and concentration at reduced pressure, isolated 6 g of oil of formula $C_6H_{10}S_2O$ by extraction with ether. This active antibacterial agent, which is more potent against *B. typhosus* than penicillin or sulfaguanidine but is otherwise less potent than penicillin, was identified as the oxide of diallyl disulfide, allyl 2-propenethiosulfinate, $CH_2{=}CH$-$CH_2S(O)SCH_2CH{=}CH_2$, termed allicin by Cavallito. Allicin, an unstable colorless liquid with an odor much more characteristic of garlic than of the various diallyl sulfides, can be synthesized by treating diallyl disulfide with oxidizing agents such as peracetic acid. Allicin is the subject of two U.S. patents in Cavallito's name, but its clinical use as an antibacterial agent was abandoned after a brief trial because of the substance's odor.

Another major contribution to understanding the chemistry of garlic was made a few years later by Swiss chemists Stoll and Seebeck of the Sandoz Company in Basel, Switzerland. In 1948 Stoll and Seebeck were able to account for the well-known fact that in the undamaged condition garlic exhibits little or no odor but, as soon as the bulbs are cut or crushed, an intense odor develops. The Swiss investigators *(27)* demonstrated that the strong-smelling unstable oil allicin is formed by enzyme cleavage from an odorless principle of higher molecular weight identified as (+)-*S*-allyl-L-cysteine sulfoxide, $CH_2{=}CHCH_2S(O)CH_2CH(NH_2)$-COOH, termed alliin. Alliin can be thought of as a molecule formed by attaching an allyl group to the sulfur of the essential amino acid cysteine followed by addition of oxygen to sulfur.

Alliin may be isolated by deep-freezing garlic bulbs in dry ice, pulverizing them while still frozen, and extracting them with ethyl alcohol containing approximately 15% water. Both the concentration of water and the temperature have to be closely regulated to avoid activating the garlic enzyme or causing it to precipitate. Solvent removal under vacuum followed by fractional crystallization from ethyl alcohol affords extremely fine, long, colorless, and odorless needles melting at 163 °C. Figure 1 summarizes the three different isolation procedures used to obtain diallyl disulfide, allicin, and alliin. Alliin, present to the extent of approximately 0.24% by weight in garlic, was the first natural product to display chirality at sulfur as well as at carbon. Alliin is readily soluble in water. Neither the characteristic odor nor the antibacterial properties of garlic are observed upon dissolving alliin in water unless the garlic enzyme allinase (itself odorless and inactive against bacteria) is also present. Under the influence of allinase, alliin is postulated to decompose to 2-propenesulfenic acid, which undergoes self-condensation, giving allicin (Figure 2). The structure of 2-propenesulfenic acid can be written with an O–H bond rather than an S–H bond inasmuch as studies by Penn

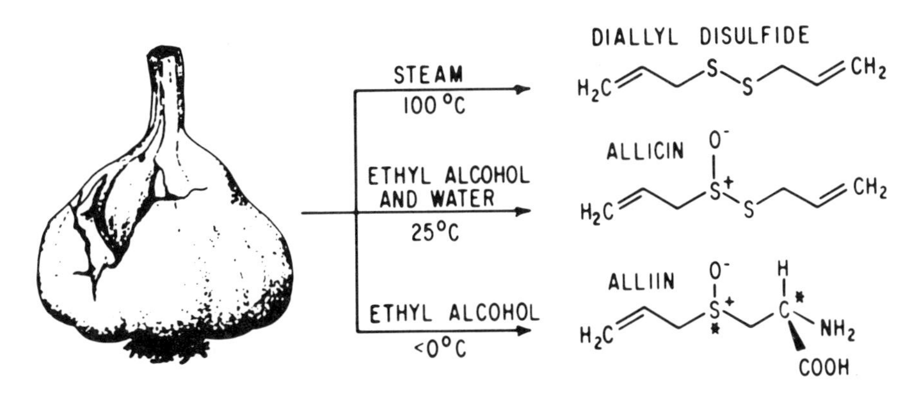

Figure 1. Different isolation procedures used to obtain diallyl disulfide, allicin, and alliin.

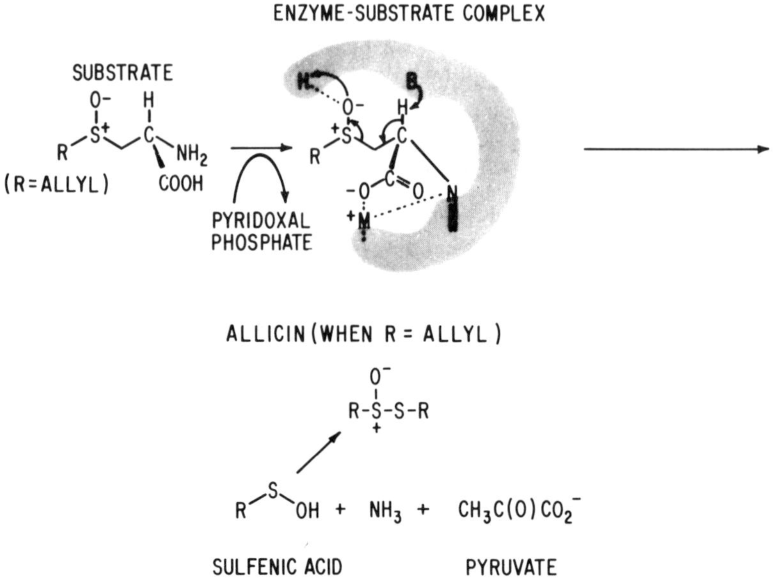

Figure 2. Decomposition of alliin to produce allicin.

and Block *(28)* have determined by microwave spectroscopy that methanesulfenic acid has the structure CH_3S–O–H. Garlic is considered to be a classic example of enzymatic development of flavor in which the initial products of reaction are highly unstable and undergo further change. Allinase has been found to act much more rapidly on the (+) optical isomer of alliin (found by X-ray crystallography to possess the *S* arrangement at sulfur) than the synthetic enantiomer with the opposite (–) or *R* configuration at sulfur *(27)*.

In 1983 Mahendra K. Jain and Roger W. Crecely of the University of Delaware, together with Rafael Apitz-Castro and Maria R. Cruz of the Venezuelan Institute for Scientific Investigations in Caracas, Venezuela, using still another extraction procedure, isolated several components from garlic that were active in preventing the aggregation of blood platelets. Their method was to soak chopped garlic pieces in aqueous methanol for 2 days at room temperature. The concentrate from this procedure was then suspended in water and extracted with ether. The ether layer was separated and concentrated and subjected to chromatographic separation. The various components were added to blood platelets that had been induced to begin aggregation with collagen or ADP. The inhibitory effects of these components on platelet aggregation could thus be determined. The most active compound was shown to have the chemical formula $C_9H_{14}S_3O$. In a collaborative effort with these researchers, Ahmad and Block *(29)* were able to establish the structure of this compound. They dubbed this new substance "ajoene", in place of the unwieldy IUPAC name 4,5,9-trithiadodeca-1,6,11-triene 9-oxide. This name comes from *ajo* (pronounced "aho"), the Spanish word for garlic.

The presence of ajoene or other antithrombotic (preventing platelet aggregation) compounds could not be detected in dehydrated garlic powder, pills, oils, extracts, or other proprietary garlic preparations. The reason is probably because most such preparations are based on the steam-distilled essential oil of garlic. These results are in accord with earlier clinical findings and with Cavallito's *(25, 26)* observation that commercial garlic oil lacks the antibacterial properties of fresh garlic.

Our earlier work on the conversion of *S*-methyl methanethiosulfinate (a lower homologue of allicin also formed from garlic) to 2,3,5-trithiahexane 5-oxide allowed us to postulate a mechanism for formation of ajoene from allicin. This example shows how ideas generated through basic research have long-range practical utility. In this case the chemical oddity lay dormant for 12 years but now allowed us to postulate an easy and efficient way of generating large quantitites of *cis*- and *trans*-ajoene for biological testing. Thus, simply heating allicin with a mixture of water and an organic solvent such as acetone afforded ajoene in approximately 40% yield *(29)*. Experiments currently under way should establish the utility, if any, of ajoene and related synthetic derivatives as drugs.

A second process occurring with allicin from natural or synthetic sources is reaction with water, giving diallyl disulfide, sulfur dioxide, and the hydrocarbon propylene. We had previously observed the decomposition of methyl methanethiosulfinate to the highly reactive thiocarbonyl compound thioformaldehyde. Evidence for the occurrence of this same type of process in the case of allicin was obtained by isolating from garlic extracts two mildly antithrombotic cyclic compounds, which we suspected were formed from allicin by way of self-condensation via Diels-Alder reactions of thioacrolein *(29)*. This latter, sapphire blue, highly reactive, unsaturated thiocarbonyl compound had already been shown by Bock *(31, 32)* at the University of Frankfurt to dimerize below -100 °C; the two cyclic compounds were found in exactly the same ratio as found by us in garlic extracts. Furthermore, when allicin was heated under high vacuum into a receiver cooled to -196 °C, the deep blue color characteristic of thioacrolein was observed in the low-temperature receptacle.

Components of garlic oil such as diallyl disulfide and methyl allyl trisulfide, which have been claimed to be the active antithrombotic factor of garlic *(33)*, were found to be virtually inactive in preventing platelet aggregation. Studies by Catalfamo of the New York State Department of Health together with the University of Delaware and Venezuela groups *(34)* suggest that ajoene acts as an antithrombotic agent by inhibiting exposure of fibrinogen receptors on platelet membranes.

Following synthesis of molar quantities of ajoene, we have attempted to determine whether this proposed mode of action is correct. One approach has involved preparation of a variety of homologues to obtain structure-activity data. As can be seen from Table I, the sulfinyl group (S=O) can be replaced by a sulfonyl group ($-SO_2-$), the disulfide group (-S-S-) can be replaced by a thiosulfinate group (-SS(O)-), and the allyl group ($CH_2=CHCH_2$) on the disulfide sulfur can be replaced with a saturated propyl group ($CH_3CH_2CH_2-$) or a crotyl group ($CH_3CH=CHCH_2-$) without loss of activity. On the other hand, saturation of the internal double bond or removal of one of the disulfide sulfurs results in loss of antithrombotic activity. On the basis of these structure-activity data as well as a variety of biochemical studies involving platelets, we postulate that platelet membranes participate in a disulfide exchange reaction with ajoene and its homologues and the alteration in the membrane prevents aggregation. We further suggest that coordination of the sulfinyl or sulfonyl group of ajoene and its homologues with calcium ions assists in the binding of these compounds to the platelet membranes.

Our hope is that use of ajoene will provide the medical benefits of garlic without the antisocial side effects. Since garlic provides a lingering reminder of the meal inasmuch as sulfur compounds are introduced into the bloodstream with slow elimination occurring

Table I. Structure-Activity Study of Antithrombotic Effect of Homologues of Ajoene

Compound Number	Structure	Structural Variation	ID_{50} (μM) ADP	ID_{50} (μM) Collagen
1	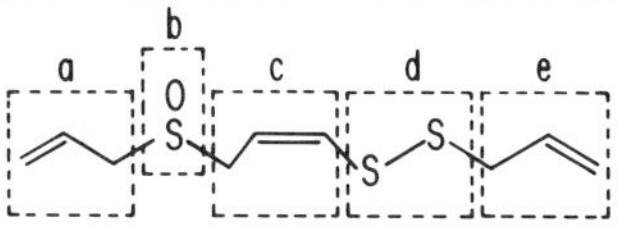 a, b, c, d, e	NONE	166±38(5)	196±64(7)
2	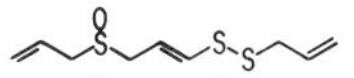	c	213	243±90(4)
3	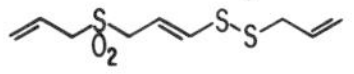	b,c	299	128±81(3)
4		b	209	214
5	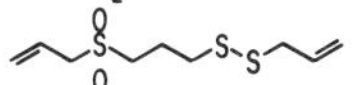	c	>400	>400
6	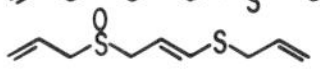	c,d	>400	>400
7	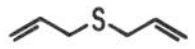	b−e	—	>400
8	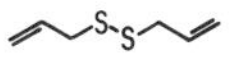	a−d	>400	>400
9	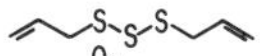	a−d	>400	>400
10	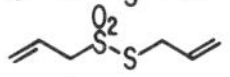	a−d	384	312
11	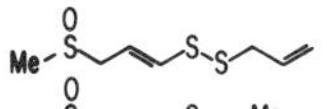	a,c	>400	374
12	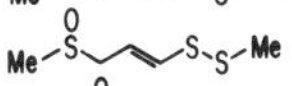	a,c,e	>400	332
13	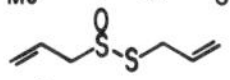	b−e	>400	400
14	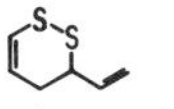	a−e	>400	>400
15	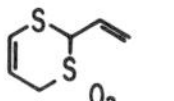	a−e	>400	>400
16	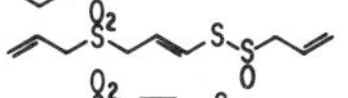	b,c,d	259	234
17	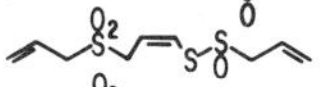	b,d	182	140
18	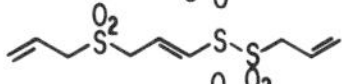	b,c,d	*	—
19	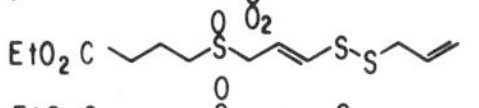	a,c	388	388
20	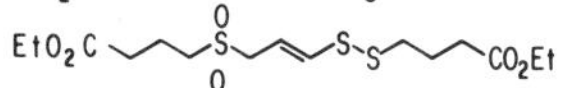	a,c,e	>400	362
21	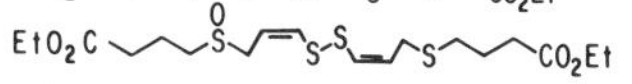	a−e	>400	>400
22	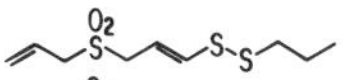	b,c,e	210	—
23	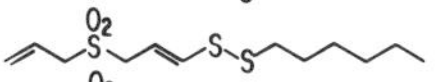	b,c,e	>400	—
24	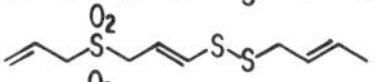	b,c,e	208	—
25	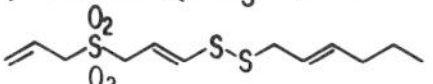	b,c,e	>400	—
26	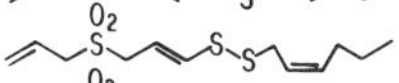	b,c,e	280	—
27	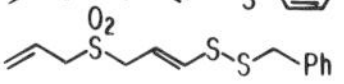	b,c,e	204	—
28	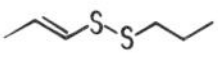	a−c,e	>>400	—

NOTE: * indicates aggregation of platelets in the absence of inductor.

in exhaled air and perspiration, therapy with fresh garlic may render the patient a social outcast for days. Even when applied to the soles of the feet, the odor of garlic may be perceived in the perspiration, breath, and urine. Of course, unless one is isolated from modern medical facilities as was Schweitzer, treatment with concoctions prepared from garlic should not be substituted for proper medical diagnosis and care.

Why did Nature incorporate in garlic the precursors to allicin? Since allicin is an antifungal agent as well as an antibiotic, allicin could offer the garlic plant protection against fungus-induced bulb decay *(35)*. Allicin and diallyl polysulfides from garlic are irritating and repugnant to certain animals ("Lilies that fester smell far worse than weeds", Shakespeare's *Sonnet 94*), providing additional survival value.

ACKNOWLEDGMENTS

I gratefully acknowledge support for this work by the donors of the Petroleum Research Fund administered by the American Chemical Society, the Northeastern New York Chapter of the American Heart Association, the John Simon Guggenheim Memorial Foundation, the Société Nationale Elf Aquitaine, and the National Science Foundation. I also acknowledge major contributions to this work on the antithrombotic factor of garlic by my collaborators, Professor Mahendra K. Jain, Dr. Rafael Apitz-Castro, Dr. James Catalfamo, and Saleem Ahmad.

LITERATURE CITED

1. Darby, W. J.; Ghalioungui, P.; Grivetti, L. "Food: The Gift of Osiris;" Academic: New York, 1977. Vol. 2, pp. 656–60.
2. Bolton, S.; Null, G.; Troetel, W. M. *Am. Pharm.* **1982,** *NS22,* 40–43.
3. Singer, M. "The Fanatic's Ecstatic Aromatic Guide to Onions, Garlic, Shallots and Leeks;" Prentice-Hall: Englewood Cliffs, N.J., 1981.
4. Harris, L. J. "The Book of Garlic;" Holt, Rinehart and Winston: New York, 1974.
5. Watanabe, T. "Garlic Therapy;" Japan Publications: Tokyo, 1974.
6. Binding, G. J. "About Garlic;" Thorsons Publishers: London, 1970.
7. Pasteur, L. *Ann. Chim. Phys. Ser.* **1858,** *52,* 404.
8. A. Schweitzer, cited by Simon, *Med. Klin. (Munich)* **1932,** *No. 3.*
9. Adetumbi, M. A.; Lau, B. H. S. *Med. Hypotheses* **1983,** *12,* 227–37.
10. Piiotrowski, G. *Praxis (Bern)* **1948,** *July 1.*

11. Lau, B. H. S.; Adetumbi, M. A.; Sanchez, A. *Nutr. Res.* **1983,** *3,* 119–28.
12. Sainani, G. S.; Desai, D. B.; Gorhe, N. H.; Natu, S. M.; Pise, D. V.; Sainani, P. G. *Indian J. Med. Res.* **1979,** *69,* 776–80.
13. Bordia, A. *Atherosclerosis (Shannon, Irel.)* **1978,** *30,* 355–60.
14. Makheja, A. N.; Vanderhoek, J. Y.; Bailey, J. N. *Lancet* **1979,** *1,* 781.
15. Makheja, A. N.; Vanderhoek, J. Y.; Bailery, J. N. *Prostaglandins Med.* **1979,** *2,* 413–24.
16. Boullin, D. J. *Lancet* **1981,** 1, 776–77.
17. Gaffen, J. D.; Tavares, I. A.; Bennett, A. *J. Pharm. Pharmacol.* **1984,** *36,* 272–74.
18. Jain, R. C. *Indian J. Med. Res.* **1976,** *74,* 1509–15.
19. Kamanna, V. S.; Chandrasekhara, N. *Lipids* **1982,** *17,* 483–88.
20. Wertheim, T. *Justus Liebigs Ann. Chem.* **1844,** *51,* 289–315.
21. Wertheim, T. *Justus Liebigs Ann. Chem.* **1845,** *55,* 297–304.
22. Liebig, J. *Justus Liebigs Ann. Chem.* **1834,** *11,* 14.
23. Zeise, W. C. *Justus Liebigs Ann. Chem.* **1834,** *11,* 1.
24. Semmler, F. W. *Arch. Pharm.* **31892,** *230,* 434–43.
25. Cavallito, C. J.; Bailey, J. H. *J. Am. Chem. Soc.* **1944,** *66,* 1950–51.
26. Cavallito, C. C.; Buxk, J. S.; Suter, C. M. *J. Am. Chem. Soc.* **1944,** *66,* 1952-54.
27. Stoll, A.; Seebeck, E. *Adv. Enzymol.* **1951,** *11,* 337-400.
28. Penn, R. E.; Block, E.; Revelle, L. K. *J. Am. Chem. Soc.* **1978,** *100,* 3622–23.
29. Block, E.; Ahmad, S.; Jain, M. K.; Crecely, R. W.; Apitz-Castro, R.; Rosa, M. R. *J. Am. Chem. Soc.* **1984,** *106,* 8295–96.
30. Block, E. *Sci. Am.* **1985,** *252,* 114–19.
31. Bock, H.; Mohmand, S.; Hirabayashi, T.; Semkow, A. *Chem. Ber.* **1982,** *115,* 1339–1348.
32. Vedejs, E.; Eberlein, T. H.; Varie, D. L. *J. Am. Chem. Soc.* **1982,** *104,* 1445–54.
33. Ariga, T.; Oshiba, S.; Tamada, T. *Lancet* **1981,** *1,* 150–51.
34. Jain, M. K.; Apitz-Castro, R.; Ledezma, E.; Vargas, J. R.; Escalante, J.; Block, E.; Ahmad, S.; Catalfamo, J., unpublished work.
35. Durbin, R. D.; Uchytil, T. F. *Phytopathol. Mediterr.* **1971,** *10,* 227–30.

0939-1/86/0137$06.00/0

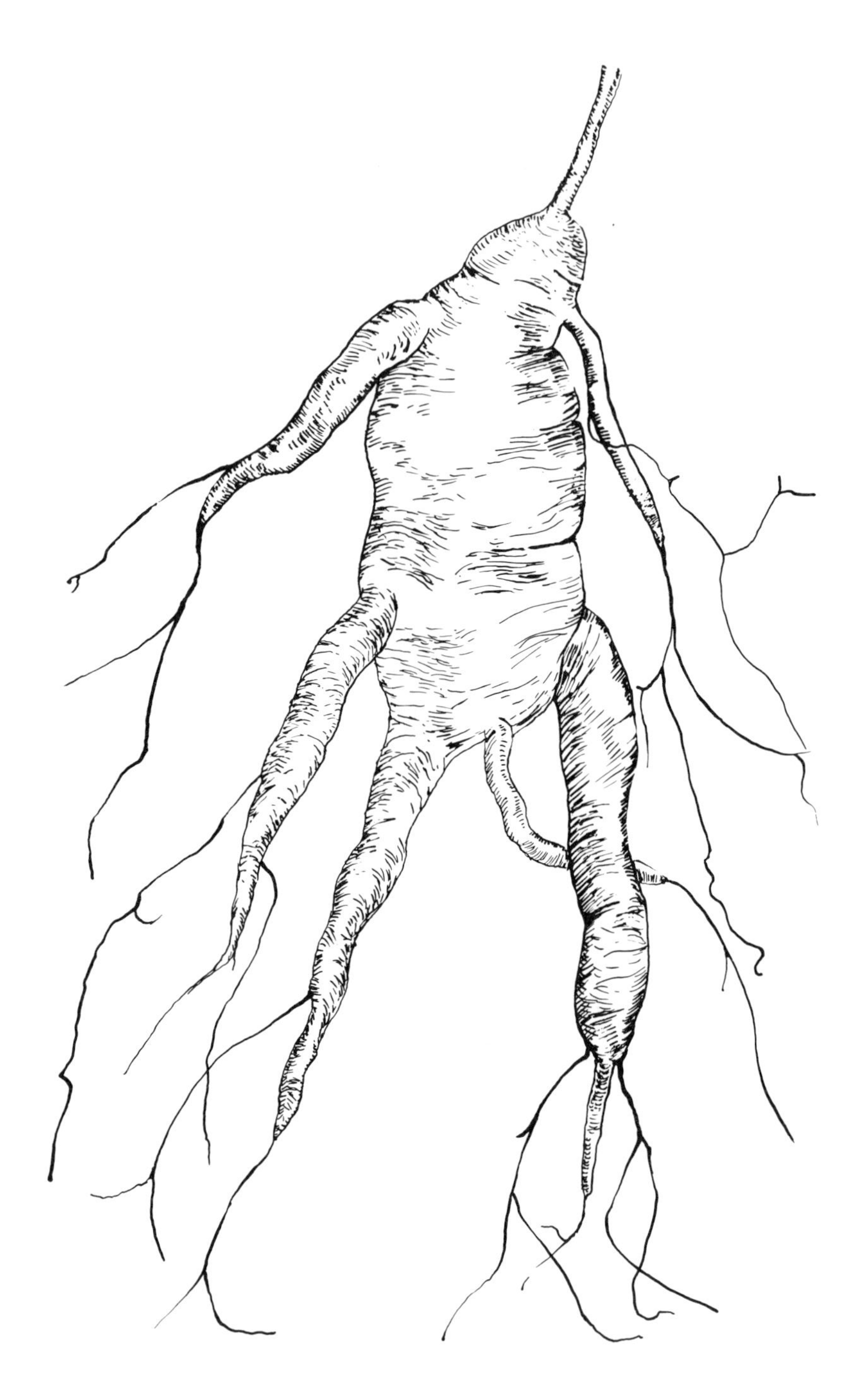

American ginseng (*Panax quinquefolium*)

Chapter 9

Scientific Basis of the Therapeutic Effects of Ginseng

T .B. NG* and H. W. YEUNG*

Ever since ancient times the root of *Panax ginseng* has been renowned for its use in traditional Chinese medicine. The medicinal plant belongs to the family Araliaceae. The plant's botanical name *P. ginseng* conveys the meaning that it is a versatile prized drug: pan means all, ax means cure, gin means man, and seng means essence. The ancient Chinese believed that the root of the ginseng plant was the crystallization of the essence of the earth in the shape of a man and that ginseng had rejuvenating, recuperative, revitalizing, and curative actions *(1)*. *Shen-nung Pen-t'sao Ching,* the first and anonymous Chinese materia medica written about 2000 years ago, stated that ginseng was used for its tonic and tranquilizing effects; that ginseng increased alertness, brilliance, and concentration, and improved memory; and that ginseng's prolonged use brought about longevity. In *Pen T'sao Kang Mu,* the famous Chinese pharmaceutical compendium authored by Li Shih-chen and first published in 1596, ginseng was described in detail *(2)*. Modern scientific findings on ginseng could be found in the Chinese pharmacopeia *(3)* and the comprehensive monograph by Oura et al. *(4)*. Proceedings of symposia also contain much information of interest on ginseng *(5, 6)*.

*Department of Biochemistry and Chinese Medicinal Material Research Center
The Chinese University of Hong Kong
Shatin, New Territories, Hong Kong

Ginseng *(P. ginseng)* grows in Northern China (Chinese ginseng) and Korea (Korean ginseng). The commercial crude drug is in either of two forms, white ginseng or red ginseng. The former is dried ginseng root and the latter is ginseng root steamed in caramel. American ginseng *(Panax quinquefolium)* occurs in the northeastern United States and eastern Canada. Japanese ginseng *(Panax japonicum)* grows in Japan. San-Ch'i ginseng *(Panax notoginseng)* and Himalayan ginseng *(Panax pseudoginseng* ssp. *himalaicus* var. *argustifolius)* are relatives of *P. ginseng.* The so-called Siberian ginseng is the Chinese drug tze-wu-chia *(Eleutherococcus senticosus,* Araliaceae) that exhibits actions similar to those of *P. ginseng* and is employed as a substitute for the much rarer *P. ginseng* in the USSR *(7, 8)*.

This review proposes to highlight some of the therapeutic effects of ginseng and to provide a scientific basis for these actions. The review is by no means an exhaustive account of the multitudinous effects of the medicinal plant. Research aimed at furnishing sound scientific evidence and explanations for the actions of *P. ginseng* would reinforce its traditional therapeutic value and dispel some of the skepticism about the plant. Much progress has been made in this direction, and approbation and credit must be given to those investigators who devote their time and energy to ginseng research. More brilliant researchers should be encouraged to participate in this fruitful area of research. Because knowledge of the chemical constituents of ginseng is imperative to the understanding of the pharmacological actions of ginseng, we shall examine the varieties of chemical constituents in different types of ginseng.

CONSTITUENTS OF DIFFERENT TYPES OF GINSENG

The types and relative abundance of the chemical constituents of ginseng depend on the species, the part of the plant, the place of origin, the method of cultivation, and the technique of extraction and probably account for the differences in therapeutic actions among different types of ginseng.

About 30 damarane-type triterpenoidal glycosides designated ginsenosides have been isolated and characterized from *P. ginseng* *(9)*. There are two types of sapogenin moieties encountered in the ginsenosides, and the structural difference between the moieties is evident from the names: protopanaxadiol and protopanaxatriol. Different ginsenosides contain different sugars although ginsenosides are all monosaccharides; the majority are glucose and the minority arabinose and rhamnose. Other compounds found in *P. ginseng* include steroidal compounds such as β-sitosterol, stigmasterol, and campesterol; sesquiterpenes such as panacene; polyacetylenic compounds such as β-elemene and panaxynol; carbohydrates; glycans; vitamins B and C; choline; peptides; amino acids; and trace elements such as manganese, iron, zinc, copper, vanadium, and germanium *(3–5, 9, 10)*.

Shibata's group *(11)* and Staba and his colleagues *(12)* contributed significantly to the understanding of the chemical constituents of American ginseng. Generally speaking, the types of saponins found in American ginseng are similar to those in ginseng; the exception is that ginsenosides R_a and R_f found in ginseng are absent from the American species. Furthermore, the relative abundance of panaxadiols and panaxatriols differs in the two species. Because American ginseng contains very few ginsenosides with central nervous system (CNS)-stimulating activity, it is regarded as a "cooling" agent and is used for the treatment of insomnia, indigestion, and toothaches. On the other hand, red ginseng, which contains ginsenosides with both stimulating and sedative activities, is regarded as a "trophic" agent and is used for its health-promoting effects and stimulating antifatigue and antistress actions in traditional Chinese medicine.

The saponins of Japanese Chikusetsu ginseng display a chromatogram different from those of white ginseng, red ginseng, and American ginseng. The Chikusetsu saponins exhibit no CNS-activating or antifatigue action. Chikusetsu saponins III–V have been isolated in crystalline form and their structures have been elucidated. Chikusetsu saponin III is made up of protopanaxadiol, glucose, and xylose. Chikusetsu saponin IV is composed of oleanolic acid, glucose, arabinose, and glucuronic acid. Chikusetsu saponin V is identical with ginsenoside R_o of *P. ginseng,* and its components are oleanolic acid, glucose, and glucuronic acid *(9, 13–16)*.

San-chi ginseng, which is the root of *Panax notoginseng,* is cultivated in the southwestern part of China. It possesses chemical constituents analogous to those of ginseng except for its higher content of ginsenosides Rb_1, R_e, and Rg_1 *(9, 17–19)*.

From Himalayan ginseng a large number of saponins have been isolated *(20–22)*. Tzu-wu-chia (Siberian ginseng) contains many saponins that resemble ginsenosides in their actions, and because of its ready availability it is used as a substitute for the much rarer *P. ginseng (8, 23–25)*.

Saponins have also been found in the terrestrial parts of ginseng including stems, leaves, flowers, buds, and fruits *(26)*, and some of the saponins are identical with the ginsenosides produced by the root, thus constituting an alternative and convenient source of ginseng saponins.

Now that we have surveyed the major chemical components of ginseng we can delve into its pharmacological actions. Novel compounds were often discovered during the course of investigating the active principle(s) for such activities, in addition to ginsenosides that display a broad spectrum of activities.

RADIOPROTECTIVE EFFECTS

Administration of an aqueous extract of ginseng to mice by either the intraperitoneal or the intravenous route prior to exposure to X-irradiation protected the animals from the damaging effects of the radiation as evidenced by an increase in the survival rate; recovery of splenic weight and DNA content; recovery of thrombocyte, erythrocyte, and leukocyte counts; and prevention of hemorrhage. Thrombocytopenia is the cause of bone marrow death in X-irradiated animals. The radioprotective effect of the ginseng extract was dose dependent *(27, 28)*. The methanol-soluble ginsenoside-containing fraction did not display any radio-protective activity. The active fraction was thermostable at pH 7 but was thermolabile under acidic or alkaline conditions. The active fraction showed ultraviolet absorption around 280 nm and exhibited a positive reaction in the biuret test and a negative reaction in the Lieberman–Burchard test for saponins, indicating that it was a protein. The radioprotective activity was not retained by (carboxymethyl)cellulose equilibrated and eluted with 20 mM KH_2PO_4 and was divisible into an unretarded peak and a retarded peak upon chromatography on Sephadex G-75 *(29)*. Although its radioprotective activity has been demonstrated in a number of experimental rodents including the mouse, the rat, and the guinea pig *(28)*, the chemical nature of the active principle remains to be elucidated.

ANTIOXIDANT ACTIVITIES

The portion of oxygen-free radicals, such as singlet oxygen, superoxide anions, and hydroxyl radicals, produced by cells during normal oxygen consumption, which escape quenching by self-protective systems in the body *(30)*, proceeds to attack unsaturated fatty acids in membranes, resulting in the generation of lipid peroxides *(31)*. The peroxide molecules that escape reduction to the less reactive hydroxy acids decompose to form malondialdehyde. Lipofuscin pigment is produced when malondialdehyde binds enzyme molecules, and the pigment increases with the course of cellular aging *(32)*. The incidence of adult diseases such as liver diseases, ophthalmic diseases, atherosclerosis, and nerve aging is also related to the accumulation of lipid peroxides. Antioxidants decrease lipid peroxidation and thus hold promise as geriatric drugs *(33, 34)*.

Han et al. *(35, 36)* studied the antioxidant activity of ginseng using acute ethanol intoxication induced lipid peroxidation in mice as a model. In the in vitro assay, linoleic acid was used as a substrate. Thiobarbituric acid, peroxide value, and activity of electron donation to 2,2-diphenyl-β-picrylhydrazyl were the measured parameters. In the in vivo assay, thiobarbituric acid, superoxide dismutase, and peroxidase activities were measured. It was discovered that antioxidant activity resided in the ether-soluble acidic and the butanol-soluble glycoside fractions but not in the water-soluble and ether-soluble neutral fractions. From the active fractions, maltol (3-hydroxy-2-methyl-γ-pyrone), salicylic acid, and vanillic acid were isolated and identified as the active compounds. However, *p*-hydroxycinnamic acid isolated from the ether-soluble acidic fraction was devoid of antioxidant activity. Maltol was detected in red ginseng but not in white ginseng, implying that maltol was formed during heat treatment of white ginseng in the production of red ginseng *(37)*. The antioxidant activity of gallic acid, a well-known antioxidant, was higher than that of maltol in vitro. However, maltol was more potent when ingested orally *(37)*. Inclusion of ferric ions in the extraction procedure diminished the antioxidant activity of red ginseng *(38)*. Because ferric ions are poorly absorbed from the gastrointestinal tract, the loss of antioxidant activity was due to a reaction between the ferric ions and the antioxidant components in ginseng. Subsequently, ten molecules of maltol were found to chelate one ferric ion. The chelation process depletes ferric ions used in the formation of the ADP–Fe^{3+} complex that is essential for the initiation of the free radical reaction *(38)* and accounts for the antioxidant activity of maltol. The chelation process also offers a scientific explanation for the traditional contraindication of decoction of ginseng in iron vessels *(39)*. The antioxidant compounds maltol, salicylic acid, and vanillic acid also exerted potent antifatigue activities as judged by their ability to prolong swimming time in mice, whereas the ginsenosides did not *(40)*.

ANTITUMOR AND ANTIVIRAL EFFECTS

A petroleum ether extract of ginseng inhibited the growth of leukemic cells LS1781, HeLa cells, and sarcoma 180 cells but not that of normal mouse embryonic cells *(41)*. Yun et al. *(42)* observed that a saponin-containing fraction and a petroleum ether extract of *P. ginseng* exerted differential inhibitory effects on DNA, RNA, and protein synthesis in sarcoma 180 cells. The saponin-containing fraction was more potent in suppressing nucleic acid than protein biosynthesis, whereas the petroleum ether extract had the opposite effect. Ginseng administration decreased the proliferation but not the incidence of pulmonary adenomas in mice induced by neonatal treatment with 9,10-dimethyl-1,2-benzanthracene. However, both the incidence and the proliferation of lung adenomas induced by urethane were abated after ginseng treatment. Ginseng treatment also decreased the incidence of lung adenomas and hepatomas in mice treated with aflatoxin B_1. Wang et al. *(43)* demonstrated that ginseng saponins increased phagocytosis of the reticuloendothelial system in normal as well as in tumor-bearing mice. The production of antibodies was also boosted. Yeung *(44)* and Yeung et al. *(45)* found that total ginseng saponin had no effect on cytotoxic T cell and natural killer cell activities in mice infected with A/WSN influenza virus. On the other hand, delayed-type hypersensitivity responses to virus and sheep erythrocytes were suppressed.

Ginseng saponins increased the adrenal cAMP level in intact but not in hypophysectomized rats, indicating that its action on the adrenal gland was an indirect one via the pituitary *(46)*. Also, ginseng saponin administration brought about an elevation of plasma levels of corticotropin and corticosterone. Various ginsenosides were capable of stimulating corticosterone production. Differences in the sugar moiety between different ginsenosides did not engender any major differences in their activities. Rather, the sapogenin moiety of the ginsenosides was indispensable for the expression of this corticosterone-stimulating activity because glycyrrhizin and saikosaponin, whose sapogenin moieties are different from those of ginsenosides, were devoid of such activity *(47)*. The activity of the ginseng saponins was not blocked by the H_1 receptor antagonist diphenhydramine, indicating that histamine was not involved *(48)*. The ability of dexamethasone, a synthetic glucocorticoid, to block the effect of ginsenosides on pituitary corticotropin and adrenal corticosterone secretion further supports the contention that the CNS–hypophyseal complex rather than the adrenal cortex is the site of action of the ginsenosides *(47)*. The binding of corticosteroids to certain brain regions in adrenalectomized rats was shown to be augmented by ginseng saponins *(48)*.

METABOLIC EFFECTS

Investigations have been conducted on the effect of *P. ginseng* on lipid metabolism. Joo *(49)* noticed that ginseng saponin administration prevented aortic atheroma formation in rabbits fed a high cholesterol diet. Abuirmeileh et al. *(50)* found that both the water-soluble and the petroleum ether soluble fractions of *P. ginseng* suppressed serum levels of total cholesterol and low-density lipoprotein cholesterol. The ratio of low-density lipoprotein cholesterol to high-density lipoprotein cholesterol decreased. (Hydroxymethyl)glutaryl coenzyme A reductase, a key enzyme in the cholesterol biosynthetic pathway, and fatty acid synthetase were suppressed by ginseng saponins. Ginseng saponins stimulated the biosynthesis of phospholipids, facilitated the transport and metabolism of cholesterol, decreased fecal excretion of bile acids and sterols after intraperitoneal administration of cholesterol, and decreased platelet adhesiveness. The ability of ginseng saponins to increase the level of high-density lipoprotein cholesterol and to stimulate lipoprotein lipase activity and hence intravascular hydrolysis of chylomicrons probably accounts for its preventive action on aortic atheroma formation *(49)*.

Ohminami et al. *(51)* demonstrated that although ginsenosides were without any effect on epinephrine-induced lipolysis in rat adipocytes, they inhibited corticotropin-induced lipolysis and insulin-stimulated lipogenesis. Adenosine, identified as a component of *P. ginseng,* manifested antilipolytic and lipogenic activities in rat adipocytes *(52)*. A peptide with antilipolytic activity has been isolated from an aqueous extract of ginseng. The peptide is characterized by a molecular weight of 1400 and by the absence of basic and aromatic amino acid residues *(53)*.

Hypoglycemic principles have been detected in *P. ginseng,* and this observation provides strong grounds for its traditional use as a diabetic remedy. Kimura's group *(54)* isolated a hypoglycemic fraction designated DPG-3-2, and Hikino et al. *(55)* isolated five hypoglycemic glycans. The hypoglycemic constituents of ginseng have recently been reviewed by Ng and Yeung *(56)*.

Ginseng saponins increased DNA and protein synthesis in rapidly dividing cells such as bone marrow cells and germ cells in the testes *(57)*.

Bombardelli et al. *(58)* illustrated the antifatigue effect of total ginseng saponins by the decrease in α-hydroxybutyrate dehydrogenase activity and the consequent reduced lactic acid production in the quadriceps muscle. Ginseng saponin treatment partially prevented the temperature decline in normal rats exposed to a cold stress (4 h at 2 °C) without affecting the plasma levels of glucose, lipid, and corticosterone. However, no similar effects were observed in adrenalectomized rats.

Huh and Choi *(59)* observed that ginseng saponins inhibited hepatic xanthine oxidase activities and decreased plasma urate levels in ethanol-induced hyperuricemic mice. However, the actions of ginseng saponins and allopurinol, a potent antigout drug, are distinctly different. Whereas allopurinol inhibited intestinal and renal xanthine oxidase activities in addition to hepatic xanthine oxidase activity, ginseng saponins had no discernible effects on the intestinal and renal enzymes. Ginseng saponins also differed from allopurinol in that the former inhibited xanthine oxidase in both control and ethanol-induced hyperuricemic animals and the latter inhibited the enzyme only in hyperuricemic animals. This example furnishes a good illustration of the prophylactic value of ginseng.

EFFECTS ON THE NERVOUS SYSTEM AND PERFORMANCE

Saito *(60)* reported that ginsenoside Rg_1 exhibited protective effects against stress-induced diminution in sexual behavior (licking, mounting, and intromission) and learning behavior (extinction and retrieval of passive avoidance response) in mice. Ginsenoside Rb_1 exerted a similar effect on sexual behavior but aggravated the decrease in learning behavior. Ginsenoside Rb_1 also potentiated nerve fiber production mediated by nerve growth factor in organ cultures of chick and mouse embryonic dorsal root and sympathetic ganglia *(61)* and also protected the neurogenic process from the inhibitory effects of colchicine and cytochalasin B although the inhibitory effects of vinblastin were left unabated. The potentiating effect of ginsenoside Rb_1 on nerve growth factor induced neurite outgrowth was independent of an increase in RNA or protein synthesis because no enhancement in the incorporation of [^{3}H]uridine and [^{3}H]leucine occurred in response to the ginsenoside. In X-irradiated mice, ginsenoside Rb_1 prevented the decline in the activity of tyrosine hydroxylase, a crucial enzyme in the catecholamine biosynthetic pathway, in the submaxillary gland. However, ginsenoside Rb_1 had no retrieving effect on the fall of the nerve growth factor level in the gland *(61)*. The nerve growth factor content and tyrosine hydroxylase activity in the murine submaxillary gland decrease with senescence. Ginsenoside Rb_1 also prevented the decline in the activity of tyrosine hydroxylase in the adrenals and hypothalami of mice *(62)* subjected to chronic hanging stress as detailed by Avakyan and Shirinian *(63)*. Thus, ginsenoside Rb_1 affects catecholamine synthesis in catecholaminergic neurons in brain, ganglia, and chromaffin cells in the adrenal medulla; nerve fiber production; and maintenance of the sympathetic nerve terminals. All these effects at least partially explain the protective effects of ginsenoside Rb_1 against stress-induced failure of memory retrieval.

The pharmacological properties of ginsenosides Rb_1 and Rg_1 have been compared and contrasted in detail by Saito and Lee *(64)*. Although ginsenoside Rb_1 possesses CNS-depressant, antipsychotic, and antiulcer activities, ginsenoside Rg_1 has exactly the opposite activities. The seemingly self-contradictory description about the effects of *P. ginseng* in *Shen-nung Pen-t'sao Ching* is thus not fallacious because both ginsenosides Rg_1 and Rb_1 are present in *P. ginseng*.

Rheoencephalic measurements indicated that ginseng treatment improved cerebral blood flow in patients with cerebrovascular deficits *(65)*. The effects of ginseng on some neurotransmitters and neuropeptides have also been studied. Tsang et al. *(66)*

provided evidence that ginseng saponins inhibited the uptake of radioactive γ-aminobutyrate (GABA), glutamate, dopamine, noradrenalin, and serotonin but not the uptake of 2-deoxy-D-glucose and leucine into rat brain synaptosomes. Because the GABA uptake was most sensitive to the inhibitory action of ginsenosides, ginseng might exert its action(s) in the central nervous system by affecting the removal of neurotransmitter substances in synaptic regions and the GABA-ergic neurons might be one of the major sites of its action. On the other hand, ginseng saponin treatment did not alter brain or pituitary levels of β-endorphin and dynorphin *(67)*.

Bittles et al. *(68)* observed that ginseng administration did not lengthen the lifespan but resulted in an exaggeration of behavioral responses in mice. Fulder *(48)* reported that behavioral response to mild stress in mice was exaggerated, whereas normal ambulatory behavior was not affected. Treatment of sportsmen aged 18–23 years with the standardized ginsenoside preparation Ginsana from Pharmaton, Ltd. (Switzerland) brought about an increase in the oxygen absorption capacity, which is an index of maximum and sustained performance capability. For the same amount of work performed, lactate production was reduced and the heart rate was slower in the absence of any changes in blood chemistry in ginseng-treated individuals *(69)*. Treatment of men and women in the age range 40–60 with Ginsana produced a beneficial effect on their mental and physical functions as judged by their reactive capacity, pulmonary function, and self-assessment tests (sleep patterns, capability, ability to concentrate, vitality, and mood) *(70)*. No difference in effectiveness between two Ginsana preparations containing 4% and 7% ginsenosides was found in animal *(71)* and human *(72)* studies. Dorling et al. *(73)* found that Ginsana shortened the time the test subjects took to react to visual and auditory stimuli and improved the respiratory quotient, which was defined as the ratio of increase in oxygen consumption during the working stage to that in the recovery stage when the subject was asked to step up and down a step in unison with a metronome. Kirchdorfer and Schmidt *(74)* treated some elderly subjects with Ginsana and observed that the treatment improved the patients' health as seen in the subjective and objective parameters measured. Ginsana treatment decreased rigidity and latent depressive feelings and increased alertness, the power of concentration, visual coordinaiton, motor coordination, and the grasp of abstract concepts *(75)*. The incorporation of ginseng into certain geriatric preparations widely distributed in many Western European countries *(74)* is thus justified.

CONCLUDING REMARKS

In the foregoing we have examined some of the pharmacological actions of ginseng and seen that some of the folkloric medicinal values of ginseng can indeed be explained by modern science. Future research on ginseng will undoubtedly lead to the discovery of novel compounds and a better understanding of the actions of ginseng. Han et al. *(76)* have already cautioned us that some of the pharmacological activities previously attributed to ginsenosides are in fact due to minor components present in the partially purified ginsenoside preparations. This observation emphasizes the importance of using pure preparations for pharmacological studies.

LITERATURE CITED

1. Hu, S. Y. *Econ. Bot.* **1976,** *30,* 11–28.
2. Li, S. C. "Pen Ts'ao Kan Mu" (Chinese Pharmaceutical Compendium) (1596); reprinted by People's Medical Publishing House: Beijing, 1977; pp. 699–710.
3. "Chung Yueh Chih" (Chinese Materia Medica); People's Medical Publishing House: Beijing, 1979; Vol. 1, pp. 1–10.
4. Oura, H.; Kumagai, A. Shibata, S.; Takagi, K. Eds. "Yakuyoninjin" (Recent Studies on Ginseng); Kyoritsu: Tokyo, 1981.
5. *Proc. 1st–4th Int. Ginseng Symp.* 1974–84; Korea Ginseng and Tobacco Research Institute: Daejeon, Korea.
6. Chang, H. M.; Yeung, H. W.; Tso, W. W.; Koo, A., Eds. "Advances in Chinese Medicinal Materials Research"; World Scientific: Singapore, 1985.
7. Li, C. P.; Li, R. C. *Am. J. Chin. Med.* **1973,** *1,* 249–61.
8. Baranov, A. I. *J. Ethnopharmacol.* **1982,** *6,* 339–53.
9. Shoji, J. In "Advances in Chinese Medicinal Materials Research"; Chang, H. M.; Yeung, H. W.; Tso, W. W.; Koo, A., Eds.; World Scientific: Singapore, 1985, pp. 455–69.
10. Fulder, S. "The Root of Being: Ginseng and the Pharmacology of Harmony"; Hutchinson: London, 1980.
11. Ando, T.; Tanaka, O.; Shibata, S. *Shoyakugaka Zasshi* **1971,** *25,* 28–32.
12. Kim, J. Y.; Staba, E. J. *Korean J. Pharmacol.* **1973,** *4,* 193–203.
13. Kondo, N.; Shoji, J. *J. Pharm. Soc. Jpn.* **1968,** *88,* 325–29.
14. Kondo, N.; Marumoto, Y.; Shoji, J. *J. Pharm. Soc. Jpn.* **1969,** *89,* 246–850.
15. Kondo, N.; Marumoto, Y.; Shoji, J. *Chem. Pharm. Bull.* **1970,** *18,* 1558–62.

16. Kondo, N.; Marumoto, Y.; Shoji, J. *Chem. Pharm. Bull.* **1971,** *19,* 1103-7.
17. Wu, M. Z. *Acta Bot. Yunnanica* **1979,** *1,* 119-24
18. Matsuura, H.; Kasai, R.; Tanaka, O.; Saruwatari, Y.; Fuwa, T.; Zhou, J. *Chem. Pharm. Bull.* **1983,** *31,* 2281-87.
19. Wei, J. X.; Wang, J. F.; Chang, L. Y.; Du, Y. C. *Acta Pharm. Sin.* **1980,** *15,* 359-64.
20. Kondo, N.; Shoji, J.; Tanaka, O. *Chem. Pharm. Bull.* **1973,** *21,* 2702-11.
21. Ibid.
22. Kondo, N.; Shoji, J. *Chem. Pharm. Bull.* **1975,** *23,* 3282-85.
23. Brekhman, I. I.; Mayansky, G. M. *Izv. Akad. Nauk. SSSR Ser. Biol.* **1965,** *5,* 762-65.
24. Brekhman, I. I.; Dardymov, I.V. *Lloydia* **1969,** *32,* 46-51.
25. Brekhman, I. I. "Eleutherococcus"; Nawka: Leningrad, 1968.
26. Tanaka, O. In "Yakuyoninjin" (Recent Studies on Ginseng); Oura, H.; Kumagai, A.; Shibata, S.; Takagi, K., Eds.; Kyoritsu: Tokyo, 1981; pp. 59-66.
27. Takeda, A. Yonezawa, M.; Katoh, N. *J. Radiat. Res.* **1981,** *22,* 323-35.
28. Takeda, A.; Katoh, N.; Yonezawa, M. *J. Radiat. Res.* **1982,** *23,*150-67.
29. Yonezawa, M.; Katoh, N.; Takeda, A. *J. Radiat. Res.* **1981,** *22,* 336-43.
30. Menzel, D. B. *Ann. Rev. Pharmacol.* **1970,** *10,* 379-94.
31. Masugi, F.; Nakamura, T. *Nutr. Food (Jpn.)* **1976,** *29,* 361-68.
32. Wolf, A.; Pappenheimer, A. M. *J. Neuropath. Exp. Neurol.* **1945,** *4,* 402.
33. Harman, D. *J. Gerontol.* **1961,** *16,* 247.
34. Harman, D. *J. Gerontol.* **1968,** *23,* 476.
35. Han, B. H.; Park, M. H.; Woo, L. K.; Woo, W. S.; Han, Y. N. *Korean Biochem. J.* **1979,** *12,* 33-40.
36. Han, B. H.; Park, M. H.; Han, Y. N. *Arch. Pharm. Res.* **1981,** *4,* 53-58.
37. Han, B. H.; Han, Y. N.; Park, M. H. In "Advances in Chinese Medicinal Material Research"; Chang, H. M.; Yeung, H. W.; Tso, W. W.; Koo, A., Eds.; World Scientific: Singapore, 1985, pp. 485-98.
38. Han, B. H. M.; Park, M. H. *Korean J. Pharmacol.* **1978,** *9,* 169-71.
39. Li, S. C. "Pen Ts'ao Kang Mu" (1596); reprinted by People's Medical Publishing House: Beijing, 1977, p. 701.
40. Han, B. H.; Park, M. H.; Han, Y. N.; Shin, S. C. *Yakhak Hoeji* **1984,** *28,* 231-35.
41. Hwang, W. I.; Cha, S. M. *Proc. 2nd Int. Ginseng Symp. (Korea)* **1978,** pp. 43-49.
42. Yun, R. S.; Lee, S. Y.; Yun, T. K. *Proc. 2nd Int. Ginseng Symp. (Korea)* **1978,** pp. 51-54.
43. Wang, B.; Cui, J.; Liu, A. In "Advances in Chinese Medicinal Material Research"; Chang, H. M.; Yeung, H. W.; Tso, W. W.; Koo, A., Eds.; World Scientific: Singapore, 1985, pp. 519-27.
44. Yeung, H. W. *Proc. 3rd Int. Ginseng Symp. (Korea)* **1980,** pp. 245-49.
45. Yeung, H. W.; Cheung, K.; Leung, K. N. *Am. J. Chin. Med.* **1982,** *10,* 44-54.
46. Hiai, S.; Sasaki, S.; Oura, H. *Planta Med.* **1979,** *37,* 15-19.
47. Hiai, S.; Yokoyama, H.; Oura, H.; Yano, S. *Endocrinol. Jpn.* **1979,** *26,* 661-65.
48. Fulder, S. J. *Am. J. Chin. Med.* **1981,** *9,* 112-18.
49. Joo, C. N. *Proc. 3rd Int. Ginseng Symp. (Korea)* **1980,** pp. 27-36.
50. Abuirmeileh, N.; Qureshi, A. A.; Din, Z. Z.; Elson, C. E.; Burger, W. C.; Ahmad, Y. *5th Asian Symp. Med. Plants Spices (Korea)* **1984,** p. 79.
51. Ohminami, H.; Kimura, Y.; Okuda, H.; Tani, T.; Arichi, S.; Hayashi, T. *Planta Med.* **1981,** *41,* 351-58.

52. Okuda, H.; Yoshida, R. *Proc. 3rd Int. Ginseng Symp. (Korea)* **1980,** pp. 53-57.
53. Ando, T.; Muraoka, T.; Yamasaki, N.; Okuda, H. *Planta Med.* **1980,** *38,* 18-23.
54. Kimura, M.; Waki, I.; Chujo, T.; Kikuchi, T.; Hiyama, C.; Yamazaki, K.; Tanaka, O. *J. Pharmacobio. Dyn.* **1981,** *4,* 410-17.
55. Hikino, H. *5th Asian Symp. Med. Plants Spices (Korea)* **1980,** pp. 229-34.
56. Ng, T. B.; Yeung, H. W. *Gen. Pharmacol.,* in press.
57. Yamamoto, M.; Uemura, T. *Proc. 3rd Int. Ginseng Symp. (Korea)* **1980,** pp. 115-19.
58. Bombardelli, E.; Cristoni, A.; Lietti, A. *Proc. 3rd Int. Ginseng Symp. (Korea)* **1980,** pp. 9-16.
59. Huh, K.; Choi, C. W. *Proc. 3rd Int. Ginseng Symp. (Korea)* **1980,** pp. 131-35.
60. Saito, H. In "Advances in Chinese Medicinal Materials Research"; Chang, H. M.; Yeung, H. W.; Tso, W. W.; Koo, A., Eds.; World Scientific: Singapore, 1985, pp. 509-18.
61. Saito, H.; Suda, K.; Schwab, M.; Thoenen, H. *Jpn. J. Pharmacol.* **1977,** *27,* 445-51.
62. Saito, H.; Nishiyama, N.; Fujimori, H.; Hinata, K.; Kamegaya, T.; Kato, Y.; Bao, T. In "Catecholamines and Stress: Recent Advances"; Usdin, F.; Kvetnansky, R.; Kopin, I. J., Eds.; Elsevier N. Holland: New York, 1980; pp. 467-80.
63. Avakyan, O. M.; Shirinian, E. A. In "Catecholamines and Stress: Recent Advances"; Usdin, F.; Kvetnansky, R.; Kopin, I. J., Eds.; Elsevier N. Holland: New York, 1980; pp. 431-35.
64. Saito, H.; Lee, Y.-M.; *Proc. 2nd Int. Ginseng Symp. (Korea)* **1978,** pp. 109-14.
65. Quiroga, H. A.; Imbriano, A. E. *Orientac. Med.* **1978,** *28,* 86-87.
66. Tsang, D.; Yeung, H. W.; Tso, W. W.; Peck, H. *Planta Med.* **1985,** *3,* 221-24.
67. Ho, W. K. K.; Ng, T. B.; Yeung, H. W.; Wen, H. L. *Biochem. Pharmacol.,* in press.
68. Bittles, A. H.; Fulder, J. J.; Grant, E. C.; Nicholls, M. R. *Gerontology* **1979,** *25,* 125-31.
69. Forgo, I.; Kirchdorfer, A. M. *Aerztl. Praxis* **1981,** *33,* 1784-86.
70. Forgo, I.; Kayasseh, L.; Staub, J. J. *Med. Welt* **1981,** *32,* 751-56.
71. Forgo, I.; Kirchdorfer, A. M. *Notabene Medici* **1982,** *12,* 721-27.
72. Kirchdorfer, A. M. In "Advances in Chinese Medical Materials Research"; Chang, H. M.; Yeung, H. W.; Tso, W. W.; Koo, A., Eds.; World Scientific: Singapore, 1985, pp. 529-42.
73. Dorling, E.; Kirchdorfer, A. M.; Rückert, K. H. *Notabene, Medici* **1980,** *10,* 241-46.
74. Kirchdorfer, A. M.; Schmidt, U. J. *Proc. 2nd Int. Ginseng Symp. (Korea)* **1978,** pp. 19-23.
75. Sandberg, F. *Sven. Farm. Tidskr.* **1980,** *84,* 199-502.
76. Han, B. H.; Park, M. H.; Han, Y. N.; Woo, L. K. *Proc. Symp. Nat. Prod. Chem. ROC-RPK* **1984,** Taipei, p. 17.

0939-1/86/0151$06.00/0

Heliotropium indicum (Reproduced with permission. Copyright 1981 Reference Publications, Inc.)

Chapter 10

Anticancer Chinese Drugs: Structure–Activity Relationships

ERIC J. LIEN* and WEN Y. LI*

Natural products hold much promise in providing valuable and novel anticancer drugs and other biologically active compounds. When many of the thousands of species of plants, including those used in either traditional or folk Chinese cancer medicine, were screened, many natural compounds with cytotoxic and/or anti-tumor activity were isolated, identified, and evaluated. Traditional Chinese medicine depends mainly on empirical approaches. For centuries experimentation was done directly on patients without going through different phases of testing in animals. As a result, thousands of herbs, plants, and preparations have been handed down to contemporary practitioners in China and many other parts of the world. Only in the last few decades have the compositions and efficacy of various Chinese herbs and plants been subjected to modern methods of analysis and screening tests. Although these compounds have quite different molecular structures and belong to various chemical groups, the presence of certain common structural features or functional groups responsi-

*Section of Biomedicinal Chemistry
School of Pharmacy
University of Southern California
Los Angeles, CA 90033

ble for their biological activity can be identified in many cases. Thus, the structure–activity relationship analysis of these active constituents is undoubtedly important either for improvement of the therapeutic effect of anticancer drugs from extracts of plants or for synthesis of novel drugs. This chapter presents a survey of chemical structures and the anticancer activities of compounds from Chinese medicinal plants as well as their mechanisms of action.

α,β-UNSATURATED CARBONYL COMPOUNDS

Sesquiterpenes

Since 1960, hundreds of sesquiterpenes with biological activities have been discovered from various plants of the Compositae family. All these compounds contain at least one conjugated carbonyl group to serve as an alkylating group. Examples are α-methylene γ-lactone, α,β-unsaturated cyclopentenone, α-methylenecyclopentanone, conjugated unsaturated ester, and other active groups *(1–4)*. These groups can react with a nucleophilic group of biological macromolecules to undergo Michael-type addition.

Besides having a high order of cytotoxicity in the KB test system in vitro (*see* Table I), a number of sesquiterpene lactones containing the α-methylene γ-lactone moiety have also been found to be potent inhibitors of Walker-256 carcinosarcoma and Ehrlich ascites tumor cells, and marginal inhibitors of P-388 lymphocytic leukemia and Lewis lung cancer *(5)*. The tumor

Table I. Cytotoxicity and Antitumor Activity of Some Sesquiterpenes

Name	*KB* ED_{50} *(μg/mL)*	*In Vivo* T/C *(%) [mg/kg]*	*Source*
Eupatolide	0.5–1.3	160[a]	*Eupatorium formosanum*
Liatrin	1.62	175[b]	*Liatris chapmanii*
Elephantopin	2.0	160 [10][b]	*Elephantopus elatus*
Molephantinin	—	397 [2.5][c]	*Elephantopus mollis*
Ambrosin	0.45	180 [35][b]	*Ambrosia maritinia*
Multigilin	—	164 [12.5][b]	*Baileya multiradiata*

[a]L-1210. [b]Ps. [c]WA.

inhibitory activities of these compounds have been proposed to be due to covalent bond formation between the O=C-C=CH_2 system and the essential SH group of key enzymes, such as deoxyribonucleic acid polymerase, thymidylase, phosphofructokinase, and glycogen synthetase. As a result of this type of Michael addition, the synthesis of DNA in tumor cells is inhibited by these compounds *(6–8)*.

Experimental data have shown that hydrogenation of or addition to these unsaturated conjugated systems by other agents resulted in a profound diminution of antitumor activity. For example, elephantopin isolated from *Elephantopus elatus* shows high cytotoxicity in the KB test system with an ED_{50} of 2 μg/mL (Table II). After hydrogenation of the double bonds conjugated with the C=O group of the lactone and that of the unsaturated ester chain ($\Delta^{11,23}$ and $\Delta^{17,18}$), the biological activity of the resulting 11,13,17,18-tetrahydroelephantopin is greatly decreased *(8)*.

A similar phenomenon has also been observed in the vernolepin series. Further evidence that the conjugated moieties are necessary for activity comes from the retention of activity after hydrogenation of the isolated $\Delta^{1,2}$ double bond of vernolepin.

Besides α-methylene γ-lactone, other conjugated systems have also been shown to be the essential functional groups for inhibition of tumor growth. Helenalin, isolated from *Helenium antuminale,* contains both an α-methylene γ-lactone moiety and an α,β-unsaturated cyclopentenone ring. Helenalin has high antitumor activities against WM-256 ascites carcinosarcoma [*T/C* = 316% at 2.5 mg/(kg · day)], P-388 (*T/C* = 200% at the 3 mg/kg

Table II. Cytotoxicity of Elephantopin and Vernolepin and Their Derivatives

Name	*KB ED_{50} (μg/mL)*
Elephantopin	2
11,13,17,18-Tetrahydroelephantopin	72
1,10,11,13,17,18-Hexahydroelephantopin	>100
Vernolepin	2
1,2-Dihydrovernolepin	2
1,2,11,13-Tetrahydrovernolepin	19
1,2,4,11,13,15-Hexahydrovernolepin	100

level), and Lewis lung cancer ($T/C = 142$ at the 25 mg/kg level) *(9–12)*.

As shown in Table III, hydrogenation of $\Delta^{2,3}$ and/or $\Delta^{11,13}$ of helenalin results in a profound decrease of cytotoxicity. This strongly suggests that the α,β-unsaturated cyclopentenone ring and α-methylene γ-lactone group contribute significantly to the biological activity, but the former apears to be more important. When the double bonds are converted into epoxy groups, the cytotoxicity of the resulting compounds only slightly decreases, indicating that the epoxyl group is also an active center. Possibly the epoxy group serves as an alkylating function reacting with the thiol group of the key enzymes involved in cell growth.

A majority of the hundreds of sesquiterpene lactones have been found to be cytotoxic and antileukemic only in vitro. A single reactive group such as α-methylene γ-lactone is not sufficient to impart significant in vivo activity. Most compounds with significant in vivo activity have been found to be multifunctional, and placement of an adjacent –OH group to conjugated C=O appears to substantially increase the potency. For example, elephantopin shows considerable activity against P-388 lymphocytic leukemia (Ps) with a *T/C* of 171% at 40 mg/kg and has been selected for further tumor panel evaluation by the National Cancer Institute (NCI) *(13)*. Deoxyelephantopin and molephantinin, isolated from *Elephantopus scaber, Elephantopus carolinianus*, and *Elephantopus mollis,* have also been shown to be active against Walker-256 carcinosarcoma in rats with *T/C* values of 226% and 397%, respectively, at the 2.5 mg/kg level *(14, 15)*.

The kinetics of the reaction between α-methylene γ-lactone and cysteine, a model biological nucleophile, have indicated that

Table III. Cytotoxicity of Helenalin and Its Derivatives

Name	*ED_{50} (HEP-2) ($\mu g/mL$)*
Helenalin	0.1
2,3-Dihydrohelenalin	3.8
11,13-Dihydrohelenalin	0.81
2,3,11,13-Tetrahydrohelenalin	40.0
2,3-Epoxyhelenalin	0.11
1,2-Epoxyhelenalin	0.53
1,2,11,13-Diepoxyhelenalin	0.50
2,3,11,13-Diepoxyhelenalin	0.50

the rate constants for these reactions at pH 7.4 range from 100 to 26,000 L mol^{-1} min^{-1}. This big difference in rates has been attributed to neighboring group effects.

Those natural products with significant in vivo activity have a rate of reaction greater than 1000 L mol^{-1} min^{-1}, and the rate of reaction of α-methylene γ-lactone with adjacent -OH or -OCOR group is generally greater than 720 L mol^{-1} min^{-1}. These facts indicate that the presence of an adjacent -OH group may enhance the electrophilicity of a conjugated system.

The compounds in which the lactone moiety does not have an adjacent -OH, or with one hydrogen at the conjugated α-methylene group replaced by alkyl, alkylamine, or alkoxy group, have decreased reactivity toward cysteine and are less cytotoxic.

Similar activation of the active functional moiety by the adjacent -OH group can also be seen in diterpenes and triterpenes.

Diterpenes and Higher Terpenes

Quassinoids

Quassinoids are degraded triterpene lactones isolated from many genera of the family Simoroubaceae. These compounds have been found to have a high degree of antileukemic activity in animals. For example, a series of bruceolides possessing significant antileukemic activities have been isolated from the Chinese drug Ya-tan-tzu, the fruit of *Brucea javanica* Merr. (*see* Table IV) *(16–19)*.

So far, among these bruceolides, bruceantin and bruceantinol are the most potent compounds. Besides their high activity against P-388 lymphocytic leukemia, these compounds also showed significant inhibitory activity against the L-1210 lymphoid leukemia and against two solid murine tumor systems, Lewis lung cancer, and B-16 melemocarcinoma. Bruceantin has been selected for phase II clinical trials *(13, 20, 21)*.

Ailanthinone and chaparrinone derivatives isolated from *Sinaba caspidata* and *Ailanthus grandis,* respectively, possess a similar basic skeleton to that of bruceolides. The dissimilarities are (1) the oxygen bridge is changed from C-8–C-13 to C-8–C-11 and (2) the hydroxyl group at C-3 is shifted to C-1. These changes do not result in a big difference either in general ring strain or in

Table IV. Antitumor Activities of Bruceolides

Name	*Activities*		*Source*
	Ps T/C *(%)* *[mg/kg]*	*KB* ED_{50} *(μg/mL)*	
Bruceoside A	156 [6]	—	*Brucea javanica*
Bruceoside B	132 [1.5]	—	*Brucea javanica*
Brusatol	198 [1]	—	*Brucea javanica*
Bruceantin	197 [0.5]	10^{-2}	*Brucea antidysenteria*
Bruceantinol	200 [0.25]	10^{-2}	*Brucea antidysenteria*
Bruceantarin	moderate	10^{-2}	*Brucea antidysenteria*
Bruceine A	moderate	—	*Brucea antidysenteria*
Bruceine B	marginal	—	*Brucea antidysenteria*
Bruceolide	marginal	—	*Brucea antidysenteria*

hydrophobic–hydrophilic balance. The presence of the conjugated ketone O=C–C=C< causes these quassinoids to also show considerable antileukemic activities *(22)* (as shown in Table V). 6α-Tigloyloxychaparrinone is effective against P-388 in vitro at concentrations less than 0.01 μg/mL, and it prolongs the life span of inflicted rats 34–63% at doses of 0.08–0.6 mg/kg.

Not surprisingly, saturation of the conjugated Δ^3 double bond or reduction of the conjugate ketone group of quassinoids is accompanied by a profound lessening in cytotoxicity. On the other hand, the neighboring hydroxyl group either at C-1 or C-3 position may enhance the reactivity of the conjugated ketone toward biological nucleophiles through formation of an intramolecular hydrogen bond.

Many naturally occurring compounds containing the unsaturated ester side chain have higher antitumor activity. For quassinoids, a 3,4-dimethyl-2-pentenoyl ester at the C-15 position in bruceolides and a senecioyl or tigloyl ester at the C-6 position in chaparrinones appear to be the most beneficial for high Ps activity and cytotoxicity. However, esterification of C-11 and C-13 hydroxy groups resulted in marked reduction of antileukemic activity, indicating that a suitable balance between hydrophobic and hydrophilic characters of the molecule is important.

Isodon *Diterpenes*

Antitumor activities of *Isodon* diterpenoids have drawn considerable attention. These compounds exist widely in plants of the

Table V. Antitumor Activities of Chaparrinone and Its Derivatives

Name	*Activities*		*Source*
	PS T/C (%) [mg/kg]	*KB* ED_{50} (μg/mL)	
Chaparrinone	145% [40]	2.0	*Sinaba multiflora*
6α-Tigloyloxy-chaparrinone	163% [6.0]	1.5×10^{-2}	*Ailanthus grundis*
6α-Tigloyloxy-chaparrin	0.24 μg/mL[a]	—	*Ailanthus grundis*

[a]This value is in vitro.

Isodon genus of the Labiatae family, for example, *Isodon amethystodes, Isodon japonicus* Hara, and *Isodon rubescens (23–25)*. Oridonin and ponicidin isolated from *I. rubescens* are the representatives of this group of compounds and have been used for the treatment of esophageal and cardiac carcinoma in clinical trials in China *(23–26)*.

Structure–activity relationship studies have shown that the α-methylenecyclopentanone moiety is the important active center for the activity of these compounds *(24, 27)*. The electrophilicity of the methylene carbon atom at the C-17 position is enhanced by formation of a hydrogen bond between the C-6–OH group and the carbonyl group at the C-15 position. This type of functional group can result in a Michael-type adduct of biological nucleophiles to the α-methylenecyclopentanone system.

Because the C-6–OH exerts an important role in polarizing the carbon atom at the C-17 position, esterification of the C-6–OH of oridonin leads to a marked decrease of antitumor activity. However, the antitumor activity of C-14–OH–acyl derivatives of oridonin increases with the increased acyl carbon chain length, which may play a carrier role in the transport process.

Tripterygium *Diterpenes*

Triptolide, which was isolated from *Tripterygium wilfordii* Hook, showed impressive life-prolonging effects ($T/C > 230\%$) in mice afflicted with L-1210 lymphoid leukemia at a 0.1 mg/kg dose level. It has been proposed *(28)* that the active center of this compound is a C-9–C-11 epoxide activated by the C-14–β-OH group through a hydrogen bond. The derivatives of triptolide lacking the C-9–C-11 epoxyl and C-14–β-OH group do not react with propanethiol (a model biological nucleophile) and are generally inactive as an antileukemic agent at doses up to 0.4 mg/kg.

Other diterpene lactones possessing antitumor activity have also been isolated from *T. wildfordii.* The mixed diterpene lactones from this plant are being clinically tested in China *(29)*.

Triterpenes and Other Compounds

Triterpenes possessing antitumor activity can be classified into three groups: (A) cardenolides, such as uzarigenin and strophanthidin; (B) withanolides, such as 4β-hydroxywithanolide E; and (C) cucurbitacins, such as cucurbitacin B. These cytotoxic and antitumor triterpenes have a steroid ring structure as well as an α,β-unsaturated lactone or ketone function.

The common characteristics of cucurbitacins are an α-hydroxy ketone in ring A and the carbonyl group at C-22 of the side chain conjugated with $\Delta^{23,24}$. Structure–activity relationship studies have indicated *(30, 31)* that the cytotoxic and antitumor activities in vivo depend on the existence of a double bond in the side chain of these compounds. Hydrogenation of the $\Delta^{23,24}$ double bond is accompanied by a profound lessening in cytotoxicity of the resultant dihydrocucurbitacin. Acetylation of the C-16–OH group of cucurbitacin B produced fabacein with marked diminution of cytotoxicity, indicating that the α,β-unsaturated ketone system is activated by hydrogen bonding between the C-16–OH group and C-22 carbonyl group.

Although cucurbitacin possesses a very high order of cytotoxicity in vitro (Table VI), the low antitumor activity in vivo and the low margins between effective and toxic doses render these materials unpromising as therapeutic agents. However, these materials may be interesting subjects for the preparation of

Table VI. Cytotoxicity of Cucurbitacins

Name	*Activity (KB ED_{50}) (μg/mL)*
Cucurbitacin B	2.5×10^{-6}
Fabecein	1.0
Dihydrocucurbitacin B	1.7×10^{-3}

semisynthetic steroidal alkylating agents with improved pharmacological properties, in view of the steroidal molecules being both lipophilic and specific as carrier moieties.

A number of quinones isolated from various plants have been shown to be active against KB cells and other tumor systems such as P-388, S-180, and WM-256.

Even though these compounds possess some structural features capable of undergoing Michael addition, none have significant in vivo activity *(4)*. This finding could be due to the lack of a proper activating group.

Podophyllotoxins (lignans of the phenyltetralin type) present in some plants of the genus *Podophyllum* (Berberidaceae) are known to possess antitumor activity through metaphase arrest. Structure–activity relationship studies have shown that the C and D rings of these molecules are involved in their interaction with tubulin; the activity of these compounds is specifically related to the configuration, size, and hydrophilic character of the C-4 substituent in ring C.

Two useful chemotherapeutic agents for the treatment of cancer have been developed by molecular modification and have been entered in clinical trials. One of them, Vp-16, is regarded as the most active compound yet tested against small cell bronchial carcinoma. The podophyllotoxin ring system is an ideal structure for further modification to discover new antitumor agents *(4)*.

ALKALOIDS

Spindle Poisons

The discovery of vinblastine and vincristine aroused a great interest in alkaloids for cancer treatment. Vinblastine is a dimeric indole alkaloid isolated from *Catharanthus roseus* (Apocyanceae).

Vinblastine (VLB) can be converted into vincristine (VCR) and vindensine (VDS) with higher antitumor activity and less toxicity by oxidation and aminolysis, respectively. The functional changes associated with the conversion of VLB to VCR and VDS generate an additional hydrogen-bonding site in the vindoline moiety and a decrease in lipophilicity in the order of VLB $>$ VDS

> VCR. Contrary to a widely held notion, VCR has higher uptake and is preferentially retained by nerve cells as well as by rat lymphoma and L-1578 cells. High intercellular levels and the low body clearance of VCR result. Differences in their action at the cellular level and in their pharmacokinetics result in differences in neurotoxicity and antitumor activity of these vinca alkaloids with a ranking order of VLB < VDS < VCR *(32, 33)*.

The biochemical mode of action is believed to involve the interaction of the vinca alkaloids with microtubular protein. This interaction alters the tertiary structure of tubulin and produces metaphase arrest of tumor cells. Other spindle poisons, for example, podophyllotoxins, colchicine, and maytansine, behave similarly. A lack of cross resistance between maytansine and VCR exists; they are different in the manner in which they inhibit microtubule assembly *(34)*.

Intercalating Agents

Camptothecine is one of the few antineoplastic alkaloids possessing consistently high activity in the resistant L-1210. In the L-1210 mouse leukemia system, camptothecine has shown activity with *T/C* values frequently in excess of 200 (Table VII). All of the naturally occurring members of the camptothecine series, including 10-hydroxycamptothecine, 10-methoxycamptothecine, and *p*-methoxycamptothecine, show the same degree of broad-spectrum activity *(13)*. The single exception is the sodium salt of camptothecine.

The structural characteristics of camptothecine are the presence of a conjugated A, B, C, D ring system and an α-hydroxy lactone moiety in ring E. Acetylation or replacement of the α-hydroxy group and reduction of the lactone ring results in the disappearance or a great reduction of activity in the L-1210 and

Table VII. Antitumor Activities of Camptothecines

Name	*Ps* T/C (%)	*L-1210* T/C (%) [mg/kg]	*9KB* ED_{50} ($\mu g/mL$)
Camptothecine	250	230 [2.0]	2×10^{-2}
10-Hydroxycamptothecine	268	230 [0.5–2.0]	2×10^{-2}

P-388 systems. These findings indicate that the α-hydroxy group adjacent to the carbonyl group plays an important role in interaction with DNA; conversion of cellular DNA to lower molecular weight fragments results. The second structural feature responsible for its activity is a flat, planar structure that is required for intercalation with nucleic acids. Although introduction of certain small substituents such as -OH, -Cl, and -OMe in ring A may result in increased activity, electric charge (e.g., $-N\equiv^{+}$) usually decreases the activity due to electronic or steric interference with intercalation with nucleic acids.

In addition to the camptothecine series, other naturally occurring compounds possessing similar conjugated planar structures include acronycine, tylocrebrine, ellipticine, indirubin, nitidine chloride, some podophyllotoxins, and some antitumor antibiotics, such as pluramycin, kidamycine, and daunomycin.

Among a wide variety of polycyclic molecules known to bind DNA by intercalation, some compounds such as actinomycin, daunomycin, and camptothecine have essential side chains or active functional groups whereas some others do not and may be considered to be "simple" intercalating agents. For example, ellipticine, present in *Ochrosia borbonica,* was found to resemble the simple intercalators, which are distinguished by a tendency for lower binding strength, faster kinetics of reaction and dissociation, and less specificity for binding to helical DNA *(35).*

Macrocyclic Compounds and Others

Maytansinoids

Maytansine isolated from *Maytenus oratus, Maytenus buchananii,* and *Maytenus hookeri* (Celastraceae) was regarded as a considerably promising anticancer agent because of its very strong inhibitory action on P-388 and other test systems (Table VIII). However, the activity of the compound in early phase II trials has been disappointing *(36).*

A maytansinoid is regarded as a benzenic ansamycin. Structure–activity relationship studies have shown that a large ester side chain at C-3 and the free hydroxy group at C-9 are the essential groups for tubulin interaction. These groups are hydro-

Table VIII. Antitumor Activity of Maytansinoids

Name	*P-388* T/C (%) [$\mu g/kg$]	*KB* ED_{50} ($\mu g/mL$)
Maytansine	220 [25]	6.1×10^{-6}
Maytanvalin	187 [12.5]	2.3×10^{-7}
Maytansinol	—	—

philic regions of the molecule in contrast to the rest of the molecule, which is hydrophobic. Compounds that lack the C-3 ester or are esterified at C-9 hydroxy group are less active than those with C-3 ester or carbinolamide with a free C-9 OH group.

Pyrrolizidine Alkaloids

Monocrotaline and indicine *N*-oxide have been shown to have significant antitumor activities *(37)*. The former, isolated from *Crotalaria sessiliflora* L, has been used effectively for the treatment of skin cancer. Indicine *N*-oxide, which exists in *Heliotropium indicum* (Boraginaceae), has been shown to have a favorable ratio of antitumor activity to toxicity and was selected for human clinical trials.

The mechanism of alkylation by dehydroretronecine responsible for the acute toxic effects and antitumor activity of pyrrolizidine alkaloids has been suggested to involve protonation and subsequent dehydration of the pyrrole; a reactive carbonium ion that is thought to be the alkylating species is formed *(38)*. The antitumor activity is most likely associated with the same functional group in the molecule responsible for hepatotoxicity.

The active pyrrolizidine alkaloids are all of the allylic ester type, which have a potential for alkylation. Their antitumor activity and hepatotoxicity are correlated with high lipid solubility and low basicity. Neither the nonester alkaloids such as rectronecine nor the ester of a saturated pyrrolizidine aminoalcohol has inhibitory action *(39)*. The importance of lipid solubilty for imparting significant in vivo activity is discussed under *Cephalotaxus* Alkaloids.

Cephalotaxus *Alkaloids*

Cephalotaxine, isolated from the plants of the *Cephalotaxus* genus, is inactive; however, the four esters listed in Table IX exhibit various degrees of antitumor activity in P-388 lymphocytic leukemia, L-1210 lymphoid leukemia, Lewis lung carcinoma, colon 38, and other experimental tumor systems *(40)*.

Clinical trials have demonstrated that harringtonine is effective in the treatment of lymphoma, Hodgkin's disease, chorriocarcinoma, primary liver cancer, and leukemia. Reportedly, the acute myelocytic leukemia has an overall remission rate of 86.1% *(41)* following treatment with this drug.

Subsequently, a series of compounds effective against P-388, such as hainanensine, hainanolide, and cephalomannine, have also been isolated from *Cephalotaxus hainanensis* and *Cephalotaxus mannii.* Cephalomannine exerts significant inhibitory activity against P-388, and an LD_{50} of 38×10^{-8} $\mu g/mL$ against KB cells. Actually, this compound is an analogue of taxol, a natural product derived from *Taxus brevifolia* and other species (Taxaceae). Taxol is a promising new antineoplastic agent; taxol has both potent antileukemic and tumor inhibitory properties and a unique mechanism of action. Rather than inhibiting tubulin polymerization, taxol is an antimitotic agent acting by promoting microtubule formations and causing shorter microtubules by decreasing the critical concentration of tubulin required for assembly. This compound should be entering clinical trials soon *(42, 43)*.

Figure 1 summarizes the biochemical mechanisms of action of various anticancer agents, including both natural products and synthetic drugs. Many drugs, such as alkylating agents, nitroso-

Table IX. Antitumor Activities of *Cephalotaxus* Alkaloids

	T/C *(%) [mg/kg]*	
Name	*P-388*	*L-1210*
Cephalotaxine	—	107 [110]
Harringtonine	405	135 [1.00]
Homoharringtonine	388	142 [1.00]
Deoxyharringtonine	180 [2.0]	—
Isoxyharringtonine	272	126 [7.5]

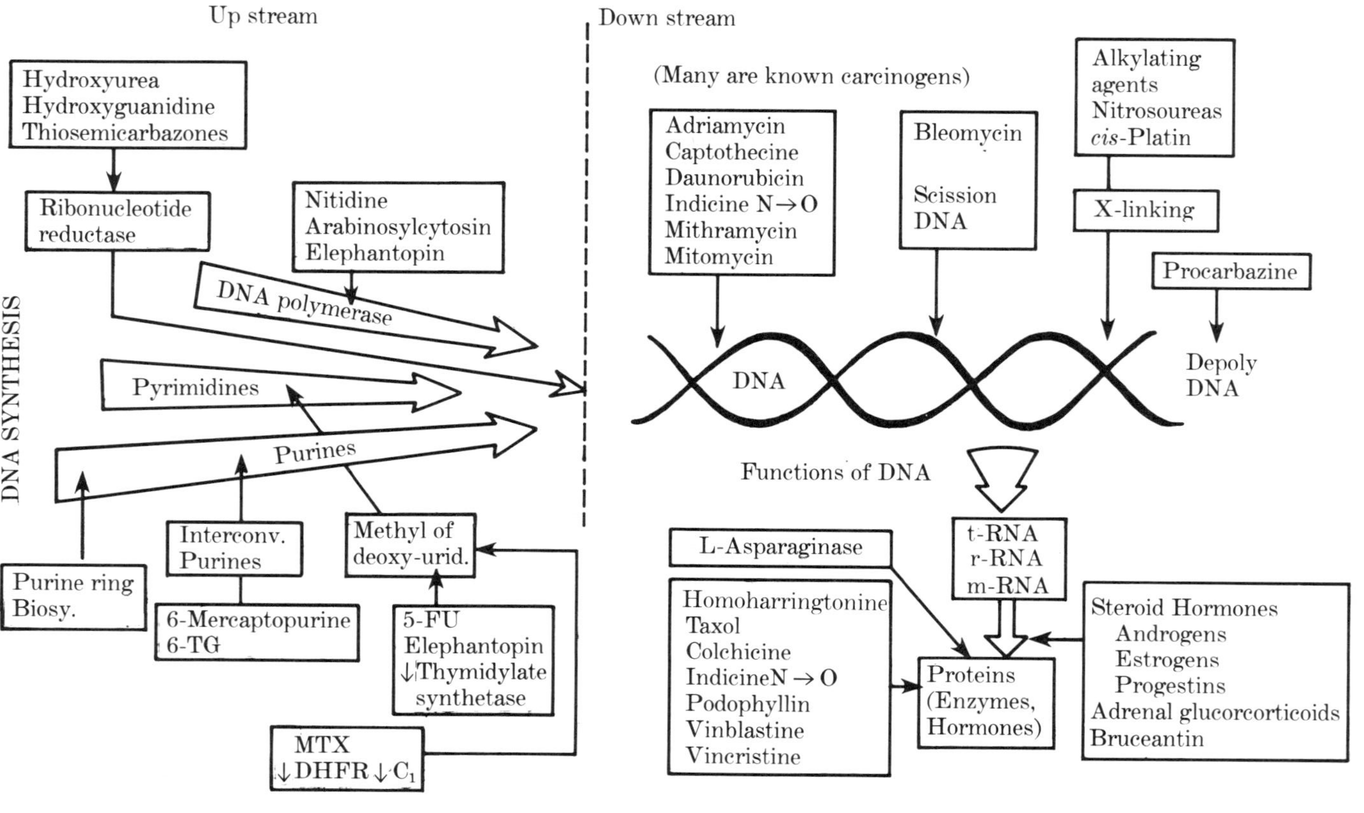

Figure 1. Mechanisms of action of anticancer drugs.

ureas, and procarbazine, that act on the preformed DNA as the target are well-known carcinogens, and the drugs acting at the upstream end of DNA synthesis are less likely to be carcinogens, unless they are converted to other reactive species *(44)*.

LITERATURE CITED

1. Zhan, B. A.; Wang, M. S. *Guewey Yaoxue, Zhiwu Yao Feng Chie* **1981,** *2(5),* 1–5.
2. Wang, Y. Z.; Zhong, S. L. *Yaoxue Xuebao* **1983,** *18(2),* 25–32.
3. Lien, E. J.; Li, W. Y. *Proc. Int. Symp. Chin. Med. Mater. Res.* Hong Kong, June 1984, in press.
4. Lien, E.J.; Li, W. Y. "Structure-Activity Relationship Analysis of Anticancer Chinese Drugs and Related Plants," Oriental Healing Hrsts. Institute: Long Beach, 1985.
5. Doskotch, R. W.; Hufford, C. D.; El-Feraly, F. S. *J. Org. Chem.* **1972,** *37,* 2740–44.
6. Smith, C. N.; Larner, J.; Thomas, A. H.; Kupchan, S. M. *Biochem. Biophys. Acta* **1972,** *276,* 94.
7. Hall, I. H.; Lion, Y. F.; Lee, K. H. *J. Pharm. Sci.* **1982,** *71,* 587–689.
8. Kupchan, S. M.; Eakin, M. A.; Giacobbe, T. J. *Science* **1970,** *168,* 376–79.
9. Lee, K. H.; Furukawa, H. *J. Med. Chem.* **1972,** *15(6),* 609–11.
10. Hall, I. H.; Lee, K. H.; Mar, E. C.; Starnes, C. O. *J. Med. Chem.* **1977,** *20(3),* 333–37.
11. Lee, K. H. *J. Pharm. Sci* **1973,** *62(6),* 1028–29.
12. Lee, K. H.; Kim, S. H.; Furukawa, H.; Piantadosi, C.; Huang, E. S. *J. Med. Chem.* **1975,** *18(1),* 59–63.
13. Beal, J. L.; Reinhard, E. "Natural Products as Medicinal Agents"; Verlag: Stuttgart, 1981; pp. 93–108.
14. Lee, K. H.; Ibuka, T.; Huang, H. C.; Harris, D. L. *J. Pharm. Sci.* **1975,** *60(6),* 1077–78.
15. Lee, K. H.; Cowhead, C. M.; Wolo, M. T. *J. Pharm. Sci.* **1975,** *64(9),* 1572–73.
16. Lee, K. H.; Imakura, Y.; Sumida, Y.; Wu, R. Y.; Hall, I. H. *J. Org. Chem.* **1979,** *44(13),* 2180–85.
17. Phillipson, J. D.; Darwish, F. A. *Planta Med.* **1981,** *41(3),* 209–20.
18. Zhang, J. S.; Lin, L. Z.; Chen, Z. L.; Xu, R. S. *Planta Med.* **1980,** *39,* 265.
19. Lin, L. Z.; Zhang, J. S.; Chen, Z. L.; Xu, R. S. *Huaxue Xuebao* **1982,** *40(a),* 75–78.
20. Kupchan, S. M.; Britton, R. W.; Lacadie, J. A.; Ziefler, M. F.; Sigel, C. W. *J. Org. Chem.* **1975,** *40,* 648–58.
21. Phillipson, J. D.; Darwish, F. A. *Planta Med.* **1979,** *35,* 308–15.
22. Wall, M. E.; Wani, M. C. *J. Med. Chem.* **1978,** *21(12),* 1186–87.
23. Wang, T. L.; Chi, C. P.; Sheng, K. H. *Zhong Caoyao* **1981,** *12(17),* 20–21.
24. Fujita, E.; Nagao, Y.; Kaneda, K.; Nakazama, S.; Kuroda, H. *Chem. Pharm. Bull.* **1976,** *24,* 2118–27.
25. Fujita, E.; Nagao, Y.; Kohno, T.; Matsuda, M.; Ozaki, M. *Chem. Pharm. Bull.* **1981,** *29,* 3208–13.
26. Chang, T. M.; Chen, Z. Y.; Chao, T. H.; Zhao, Q. Z.; Sun, D. D.; Lin, Z. W. *K'o Hsueh T'ung Pao* **1981,** *25,* 1051–55.
27. Fujita, E.; Nagao, Y.; Node, N.; Kaneko, K.; Nakazawa, I.; Kuroda, H. *Experienta* **1976,** *32,* 203–8.

28. Kupchan, S. M.; Schubert, R. M. *Science* **1974,** *185,* 791.
29. Chu, D. Y. *Zhong Caoyao* **1981,** *13(8),* 41-48.
30. Kupchan, S. M.; Tsou, G.; Sigel, C. W. *J. Org. Chem.* **1973,** *38,* 1420-21.
31. Konopa, J.; Matuszkiewicz, A.; Hrabowska, M.; Onoazka, K. *Arzneim. Forsch.* **1974,** *24,* 1741-43.
32. Told, G. C.; Griffing, W. J.; Gibson, W. R.; Morton, D. M. *Cancer Treat. Rep.* **1979,** *63(1),* 35-41.
33. Valdivieso, M.; Bedikian, A. Y.; Bodey, G. P.; Freireich, E. J. *Cancer Treat Rep.* **1981,** *65,* 877-79.
34. Ludena, R. F.; Roach, M. C. *Biochemistry* **1981,** *20(19),* 4444-50.
35. Kohn, K. W.; Waring, M. J.; Laubiger, D. G.; Friedman, C. A. *Cancer Res.* **1975,** *35,* 71-76.
36. Pinedo, H. M., "Cancer Chemotherapy, Annual 3," Elsevier: New York, 1981; p. 142-43.
37. Li, W. Y.; Lien, E. J. *Bull. Orient. Healing Arts. Inst. USA* **1984,** *9(7),* 321-32.
38. Robertson, K. A. *Cancer Res.* **1982,** *42,* 8-13.
39. Culvenor, C. C. J. *J. Pharm. Sci.* **1968,** *57,* 1112-17.
40. Powell, R. G.; Weisleder, D.; Smith, C. R. Jr. *J. Pharm. Sci.* **1972,** *61(8),* 1227-30.
41. Ma, G. E.; Lu, C. N.; Fan, G. J. *Yaoxue Tongbao* **1982,** *17(4),* 205-6.
42. Hamel, E.; Lin, C. M.; Johns, D. G. *Cancer Treat. Rep.* **1982,** *66(6),* 1381-86.
43. Schiff, P. B.; Fant, J.; Horwitz, S. B. *Nature* **1979,** *277,* 665-67.
44. Lien, E. J.; Ou, X C. *J. Clin. Hosp. Pharm.,* in press.

0939-1/86/0168$06.00/0

Chapter 11

Some Recent Biochemical Characterizations of Chinese Herbal Preparations

T. Y. SHEN*, M. N. CHANG*, S. B. HWANG*, L. Y. HUANG*, R. W. BURG*, Y. K. LAM*, G. ALBERS-SCHONBERG*, V. J. LOTTI*, WANG XU†‡ and G. Q. HAN†

A current trend in new drug development is to put emphasis on the biochemical mechanism of drug action. The premise is that a drug molecule with a well-chosen and specific mechanism of action is more likely to achieve clinical efficacy and to minimize nonspecific or unexpected side effects. The implementation of such a rational approach to drug discovery is greatly facilitated by recent advances in biological sciences, especially in biochemical pharmacology, which has elucidated many enzymatic pathways and receptor mechanisms involved in the pathogenesis of various clinical disorders.

In place of conventional animal models, in vitro biochemical assays have become high-capacity primary screening assays for potential therapeutic agents. These biochemical assays are highly specific and exquisitely sensitive. The successes of H2-receptor antagonists for ulcer treatment *(1, 2)* and angiotensin I converting

*Merck Sharp & Dohme Research Laboratories
Rahway, NJ 07065
†Beijing Medical College
Beijing, People's Republic of China
‡Deceased

enzyme inhibitors for hypertension *(2–4)* are two of the most recent examples of magic biochemical bullets. In some cases, judicial combinations of several mechanism-specific agents may be used to achieve maximum efficacy or to counterbalance side effects in a manner reminiscent of the multicomponent or multispecies mixture of traditional medicine. Many enzyme or receptor assays can also detect the presence of a minute amount of an active substance in a crude mixture, analogous to the discovery of antibiotics in fermentation broths. These biochemical assays may be applicable in the study of traditional herbal medicine as well.

In the past 3 years, a collaborative effort between Beijing Medical College and Merck Sharpe & Dohme Research Laboratories has produced a broad biochemical evaluation of a selection of Chinese herbal extracts. The isolation and identification of active ingredients from crude aqueous or methylene chloride extracts are much facilitated by the capacity, sensitivity, and specificity of the in vitro assays and by the efficiency of modern isolation and structure determination techniques. This study might correlate in vitro findings with known pharmacological and clinical effects and elucidate the principal biochemical mechanism of action of some traditional herbal preparations. In a broader sense, the Beijing Medical College is interested in correlating the biochemical profile with the clinical usage of herbal plants. The screening data may provide some insight to a scientific interpretation of certain traditional concepts, namely, the use of tonics *(bu-qi-yao,* replenishing the vital energy) and nutrients *(bu-xue-yao,* nourishing the blood) in Chinese herbal medicine.

In addition, the discovery of any minor ingredient with a novel chemical structure or bioactivity, which may be unrelated to the traditional usage of the herbal plant, is also of interest. These structures may provide medicinal chemists with a new lead for further structural modification and improvement of its therapeutic activities.

From the onset we recognized that, in our biochemical analysis of herbal plants, we may encounter several hypothetical situations (Table I). For example, Substance 1 may exert its cardiovascular effect in humans by the biochemical mechanism A, that is, by a direct in vitro and in vivo correspondence. Next, the biochemical activity (B) of Substance 2 may be related to its clinical utility in

Table I. Analysis of Herbal Plants

Chemical Substance	*Biochemical Activity*	*Clinical Usage*
1	A	cardiovascular–renal
2	B	cardiovascular–renal, pulmonary
3	C	central nervous system
	D	inflammation, pain, immunological
4	?	immunological
5	E	?

both cardiovascular and pulmonary disorders; that is, some aspects of the two diseases share a common biochemical mechanism. In another situation, a less specific Substance 3 may possess two major biochemical activities (C and D) that contribute to its broad clinical utility in several therapeutic areas. Furthermore, because the properties of immunological mediators, for example, interleukens and cytokines, and the nature of specific cellular responses in the immune network are still being elucidated, the immunological effects of Substance 4 may be demonstrable in current cellular or animal models of immune responses, but no definitive biochemical explanations can yet be provided. Finally, Substance 5 may turn out to be a novel regulator of a newly recognized enzyme or mediator whose clinical relevance has yet to be established.

Some of these situations have indeed been encountered in our study of the crude extracts from approximately 200 Chinese herbal species. We will briefly describe a few specific examples. In particular, some details in the discovery and characterization of kadsurenone as a specific receptor antagonist of a newly identified lipid inflammatory mediator, the platelet activating factor (PAF), are presented to illustrate certain aspects of the laboratory process in new drug development.

Results from four membrane receptors, treated with a variety of biochemical assays, are illustrative of our methodology. These are the α-adrenergic receptor *(5)* for biogenic amines, the benzodiazepine receptor *(6, 7)* for agents affecting the central nervous system, the cholecystokinin (CCK) receptor for neuropeptides *(8)*, and the PAF receptor for an inflammatory phospholipid mediator *(9)*. These receptors interact with distinct chemical types of ligands to produce very different biological responses.

CCK is a peptide containing 33 amino acids. CCK-8, Asp-Tyr(SO_3H)-Met-Gly-Trp-Asp-Phe-NH_2, the C-terminal octapeptide of CCK, and related peptides function as neurotransmitters and central satiety factors in appetite control *(10)*. Previously, two synthetic structures, proglumide and benzotript *(11)*, have been found to be moderately potent CCK receptor antagonists active at micromolar concentrations in vitro. Both compounds are hydrophobic carboxylic acid derivatives. From the methylene chloride extract of *Ziziphus spinosa* (Suanzaoren), we found that a lipophilic and acidic ingredient, ursolic acid ($IC_{50} = 39.4 \times 10^{-6}$ M), blocks the binding of the ligand CCK-8 to the receptor preparation but in a nonspecific manner. Several fatty acids and triglycerides from other herbal extracts also behave similarly *(12)*.

The α-adrenergic receptors have been studied extensively. When [^{3}H]clonidine and [^{3}H]WB 4101 (α-antagonist) were used as ligands for the α_2 and α_1 receptors *(13)*, respectively, two flavone derivatives, targeretin and nobiletin, isolated from *Citrus aurantium* (zhishi) extract, were found to interact with the α_2 receptor with IC_{50} values of 2.69×10^{-6} M and 2.49×10^{-6} M, respectively. Interestingly, nobiletin also showed a selectivity for competing with α_2 over α_1 receptors. Its selectivity ratio of 150 compares favorably with the selectivity ratios of several synthetic drugs of this class.

A variety of structures have been shown to compete with [^{3}H]diazepam in binding to the benzodiazepine receptor prepared from rat cerebral cortex membrane. Alantolactone ($IC_{50} = 86.2 \times 10^{-6}$ M) and isoalantolactone from the root of *Aristolochia debilis* (qingmuxiang), which are anthelmintics, were also active on the benzodiazepine receptor assay. Medicarpin, isolated from the roots of *Hedysarum polybotrys* (hongqi), has an IC_{50} of 32.2×10^{-6} M in the benzodiazepine receptor assay. It also delayed the onset of pentylenetetrazole-induced convulsions in mice when administered intraperitoneally at 200 mg/kg *(12)*.

The affinity of these active structures for receptors is relatively weak. Micromolar concentrations are needed to compete with nanomoles of ligands. However, the structures do provide some additional structure–activity information in the search for novel antagonists of these pharmacological receptors. A more potent and promising antagonist was discovered in our study of the receptor for PAF. PAF is a highly potent, endogenous phospholipid molecule whose structure was elucidated only 5 years ago *(14)*. PAF is released by inflammatory cells, such as leukocytes, basophils, and platelets. From stimulated macrophages PAF is released along with prostaglandins, leukotrienes, and interleuken 1. Elevated levels of PAF have been found in the urine of lupus patients, in the plasma of exercise-induced bronchoasthmatics, and in several laboratory models of immune-complex diseases and anaphylaxis. Compared with other well-established mediators of inflammation and anaphylaxis, PAF affects a wide range of target cells or tissues *(15)*. Administration of very small doses of exogenous PAF, as low as 10^{-12} M in vitro and 10 ng/kg in vivo, induces significant cellular and vascular changes, such as the aggregation and degranulation of platelets and neutrophils, an increase of vascular permeability, and bronchoconstriction. In human volunteers, intradermal injection of 1-60 ng of PAF in the forearm produces an immediate vasoconstriction, burning pain and itching, and a wheal and flare reaction. The vasoconstricting effect of PAF is estimated to be 1000 times more intense than that of histamine. A marked synergistic effect of PAF and prostaglandin E was also observed.

In an attempt to regulate this new lipid mediator, we established a receptor binding assay *(9)* by using isolated rabbit platelet plasma membrane and tritiated PAF to search for PAF receptor antagonists. Obviously, a specific antagonist should bind to the receptor protein but exert no agonistic activities.

Haifenteng is a Chinese herbal perparation of *Piper futokadsurae* for the general relief of bronchoasthma and the stiffness, inflammation, and pain of rheumatic conditions. We found that a methylene chloride extract of haifenteng supplied by Beijing Medical College contained a highly potent and orally active receptor antagonist of PAF ($IC_{50} = 1 \times 10^{-7}$ M). This active ingredient was isolated and identified as kadsurenone, a pseudolignan derivative *(16)*. Kadsurenone is a competitive inhibitor of the

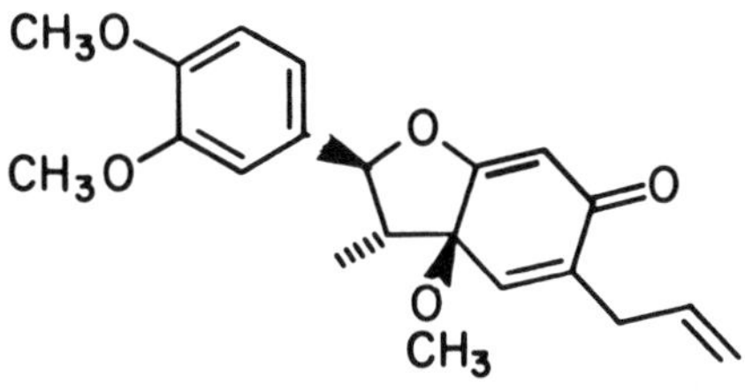

CH_3O CH_3O O O O CH_3

Kadsurenone

specific binding of tritiated PAF to its receptor preparation with a K_I of 3.88×10^{-8} M, which is less than 10 times higher than the K_I for PAF itself (6.3×10^{-9} M).

At the cellular level, kadsurenone inhibits the aggregation of rabbit platelets induced by PAF without showing any agonistic activity even at high levels. Kadsurenone also inhibits the aggregation and degranulation of guinea pig and human neutrophils at 2–5 μM. The biological specificity of kadsurenone was ascertained by its lack of effect on a variety of cellular and tissue responses to other mediators and a selection of receptor and enzyme preparations. For instance, kadsurenone has no effect on the aggregation and degranulation of platelets and human neutrophils induced by different mediators. The chemical specificity of kadsurenone was demonstrated by the very weak activity of several closely related analogues, piperenone and kadsurins A and B, isolated from the same plant *(17)*. These analogues have similar polar substituents and, as determined by X-ray crystallography, molecular stereochemistry, but they are 10–100 times less effective than kadsurenone as PAF receptor antagonists.

In vivo kadsurenone inhibits PAF-induced cutaneous vascular permeability changes in the guinea pig in a dose-dependent manner. Kadsurenone also inhibits PAF-induced hypotension, systemic vascular permeability increase, and release of lysosomal enzymes in the rat at 20–50 mg/kg *(18)*. In follow-up studies in our laboratories and by other investigators, kadsurenone appears to be an effective PAF receptor antagonist in several mechanism-based in vivo models in the rat, guinea pig, and rabbit. Work is in progress to demonstrate its effect in various disease models such as endotoxin-induced shock in rats *(19)*, in order to delineate its potential clinical applications.

More recently, another PAF-antagonist of Chinese herbal origin, ginkgolide B (BN 52021), was isolated from extracts of *Ginkgo biloba* leaves in the I. H. B. Laboratories *(20)*. Ginkgolide B has an IC_{50} of 3.6×10^{-7} M in our membrane receptor assay and is fully effective in various in vivo animal models.

In conclusion, we have demonstrated that specific biochemical assays can be used effectively to detect active ingredients in crude extracts and to profile the biochemical properties of herbal plants. Two traditional Chinese remedies have now provided two novel and potent receptor antagonists for a lipid mediator whose structure was elucidated only 5 years ago. We are optimistic that the application of new biochemical assays to natural product research will continue to be a highly efficient and productive approach in the current renaissance of folk medicine studies.

LITERATURE CITED

1. Durant, G. J.; Emmet, J. C.; Ganellin, C .R.; Miles, P. D.; Prain, H. D.; Parsons, M. E.; White, G. R. *J. Med. Chem.* **1977,** *20,* 901.
2. Brogden, R. N.; Heel, R. C.; Speight, T. M.; Avery, G. S. *Drugs* **1978,** *15,* 93.
3. Ondetti, M. A.; Rubin, B.; Cushman, D. W. *Science* **1977,** *196,* 441.
4. Horovitz, Z. P. In "Pharmacological and Biochemical Properties of Drug Substances"; Goldberg, M. E., Ed.; Am. Pharm. Assoc.: Washington, D. C., 1981; Vol. 3, p. 148.
5. Uprichard, D. C.; Snyder, S. H. *J. Biol. Chem.* **1977,** *252,* 6450.
6. Costa, E.; Guidotti, A. *Annu. Rev. Pharmacol. Toxicol.* **1979,** *19,* 531.
7. Squires, R. F. *Nature* **1977,** *266,* 732.
8. Chang, R. S. L.; Lott, V. J.; Martin, G. E.; Chen, T. B. *Life Sci.* **1983,** *32,* 871.
9. Hwang, S.-B.; Lee, C.-S.C.; Cheah, M. J.; Shen, T. Y. *Biochemistry* **1983,** *22,* 4756.
10. Smith, G. P.; Jerome, C.; Cushiw, E. R.; Simansky, K. J. *Science* **1981,** *213,* 1036.
11. Hahne, W. F.; Jensen, R. T.; Lemp, G. F.; Gardener, J. D. *PNAS (USA)* **1981,** *78,* 6304.
12. Lam, Y.-K. T.; Lott, V. J.; Chang, R. S. L., private communication.
13. Greenberg, D. A.; Uprichard, D. C.; Snyder, S. H. *Life Sci.* **1976,** *19,* 69.
14. Vargafig, B. B.; Benveniste, J. *Trends Pharmacol. Sci.* **1983,** *4,* 341.
15. Benveniste, J.; Arnoux, B., Eds. "Platelet-Activating Factor and Structurally Related Ether-Lipids"; Elsevier: Amsterdam, 1983; INSERM Symp. 23.
16. Shen, T. Y.; Hwang, S.-B.; Chang, M. N.; Doebber, T. W.; Lam, M.-H. T.; Wu, M. S.; Wang, X.; Han, G. Q.; Li, R. Z. *PNAS (USA)* **1985,** *82,* 672.
17. Chang, M. N.; Han, G. Q.; Arison, B. H.; Springer, J. P.; Hwang, S.-B.; Shen, T. Y. *Phytochemistry,* in press.
18. Doebber, T. W.; Wu, M. S.; Shen, T. Y. *Biochem. Biophys. Res. Commun.* **1984,** *125,* 980.
19. Doebber, T. W.; Wu, M. S.; Robbins, J. C.; MaChoy, B.; Chang, M. N.; Shen, T. Y. *Biochem. Biophys. Res. Commun.* **1985,** *127,* 799.
20. Braquet, P.; Spinnewyn, B.; Bourgain, R.; Drieu, K.; Defeudis, F. V. Conf. Prostaglandins, Kyoto, Japan, Nov. 1984, Abstract 013-3.

0939-1/86/0175$06.00/0

Chapter 12

Bioactive Compounds from Three Chinese Medicinal Plants

REN-SHENG XU*

China is a country rich in medicinal plant resources and traditional experiences. In recent years hundreds of Chinese scientists have focused their attention on the herbs used in folk medicine and those described in ancient classical pharmacopeias. As a result, many significant papers have been published in chemistry and pharmacology journals worldwide. In this chapter recent research results on new antiinflammatory compounds isolated from three Chinese medicinal plants are described. Structure elucidation, bioactivities, and the relationship between structure and bioactivity will be discussed.

SARMENTOSIN AND ISOSARMENTOSIN

Chui-pen-tsao, the whole plant of *Sedum sarmentosium* Bunge (Crassuaceae), is used in folk medicine in the Shanghai district for treatment of hepatitis *(1, 2)*. In clinical trials patients suffering from chronic viral hepatitis were given a decoction of the plant. Researchers observed a significant lowering of the serum glu-

*Shanghai Institute of Materia Medica
Academia Sinica
Shanghai 200031, China

tamic-pyruvic transminase (SGPT) level. This result encouraged workers in the field to search for the active compound(s) in the plant. However, no such compounds could be found when animals with tetrachloromethane-, thioacetamide-, or galactosamine-induced hepatitis were used to screen the various fractions isolated from chui-pen-tsao. For this reason, double-blind, well-controlled trials were performed on human hepatitis patients. All isolated materials were first administered to animals to guarantee the safety of the people involved. This approach successfully identified an active compound, which was called sarmentosin.

The initial extraction was done with hot ethanol. This extract was concentrated and partitioned between chloroform and water. The aqueous layer showed activity. This solution was concentrated and chromatographed on an active charcoal column with water and 70% ethanol. The ethanol eluents were combined, concentrated, and chromatographed on a silica gel column with ethyl acetate and ethyl acetate–methanol. The final sarmentosin fractions were combined and again chromatographed twice on silica gel with ethyl acetate–acetone and chloroform–methanol. The active compound, pure sarmentosin, was finally obtained as a colorless syrup that dissolved easily in water (yield about 0.1%): $[\alpha]_D$ −17.4° (c 0.62, H_2O).

The molecular formula of sarmentosin was determined by mass spectrometry (MS) and elemental analyses as $C_{11}H_{17}NO_7$: M^+ m/z 276 (M + 1). Proton nuclear magnetic resonance (^{1}H-NMR) spectral analyses show sarmentosin has an unsaturated nitrile group and a sugar component in its molecule. Other spectral data are as follows: UV 212 nm (log ϵ 4.07); IR 3540–3240 (-OH), 2235 (-CN), 1640 (-C=C-) cm^{-1}. Emulsion hydrolysis of sarmentosin led to a glucose but no aglycon moiety due to its lability. The ^{1}H-NMR spectrum of its pentaacetate, $C_{21}H_{27}NO_{12}$ (mp 79–80 °C), shows the presence of five acetyl signals at δ 1.91, 1.96, 1.96, 2.01, and 2.07; an anomeric proton of glucose at δ 4.57 (1 H, d, $J = 7$ Hz); and an olefinic proton at δ 6.54 (1 H, t, $J = 7$ Hz). The other signals are determined by decoupling techniques. When deuterium benzene was used as the solvent for measuring the ^{1}H-NMR spectrum, the olefinic proton signal shifted upfield (Δ +0.63) more than that of -O-CH_2- linked to glucosyl moiety (Δ +0.56). This indicates that the nitrile group is located trans to the olefinic proton. Carbon-13 nuclear magnetic resonance (^{13}C-NMR) spectral analysis also coincides with the proposed structure.

When pentaacetylsarmentosin was treated with barium methanolate, sarmentosin could be recovered quantitatively. Treatment of sarmentosin with dilute alkali at room temperature gave a new crystalline material, isosarmentosin, with the same molecular formula: mp 211–12 °C; $[\alpha]_D$ 51.4° (*c* 0.1, H_2O); lack of UV absorption; IR 3460, 3270 (-OH), 2230 (-CN) cm^{-1}. The olefinic proton signal in the ^1H-NMR spectrum has disappeared, and two signals at δ 3.20 (1 H, m) and 3.48 (1 H, m) are now observed. Isosarmentosin is not hydrolyzed by either acid or emulsin. These results lead us to believe that an internal Michael-type reaction occurs.

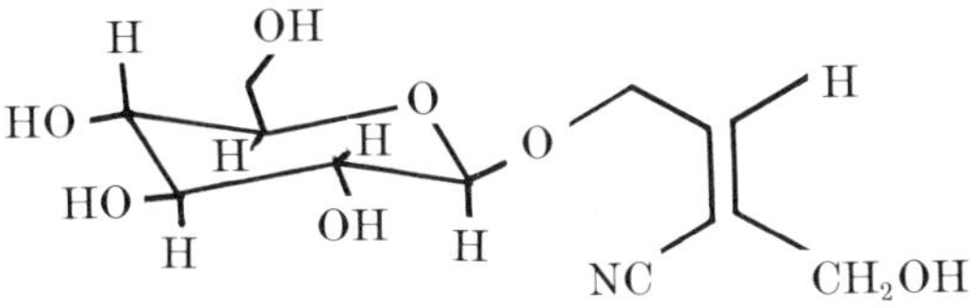

sarmentosin

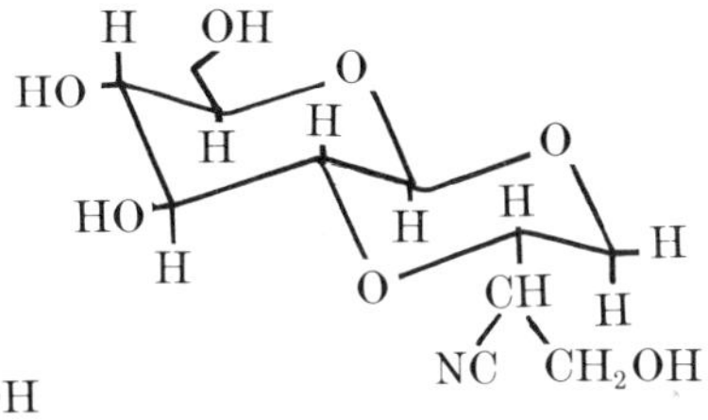

isosarmentosin

The ^{13}C-NMR spectral data agree with the proposed structure. Isosarmentosin is not active in clinical trials with chronic hepatitis patients. Structures for sarmentosin and isosarmentosin were confirmed by X-ray diffraction studies on crystalline pentaacetylsarmentosin and on isosarmentosin *(3, 4)*. These compounds belong to a new type of cyanogenic glucosides not reported previously in the literature. Simultaneous with our report, Nahrstedt isolated sarmentosin epoxide from *Sedum cepaea*. Nahrstedt *(5)* suggested that isoleucine, a precursor to other cyanogenic glucosides, could be involved in the biogenetic pathway to these compounds.

Pharmacological studies show sarmentosin possesses a suppressive effect on cell-mediated immune responses in mice. Sarmentosin can significantly decrease the number of thymus cells with almost no change on thymus and spleen weight. Sarmentosin can also decrease both the number of plaque-forming cells of the mouse spleen against sheep red blood cells and the graft versus host reaction of rats. After successive injections, the white blood cell counts and also the percentage of neutrophilic cells in the peripheral blood were increased, and the percentage of the α-naphthyl acidic esterase positive cells (T cells) was increased from 3.5 to 9.0% in the bone marrow (Tables I–IV) *(6)*. These results lead to the supposition that sarmentosin behaves like steroid hormones; that is, sarmentosin can transfer peripheral T cells into the bone marrow. This transfer would explain why hepatitis patients who stop using sarmentosin again experience a rise in their SGPT level.

Table I. Effect of Sarmentosin on Mouse Immune Organs

Items	*Expt. No.*	*Control*	*Sarmentosin*	*S/C[a] × 100*
Peripheral WBC[b]	1	84 ± 15	113 ± 20	135
($\times 10^2$/mm^3)	2	98 ± 20	141 ± 16	144
Neutrocytes	1	23 ± 10	28 ± 3	122
(per 100 WBC)	2	19 ± 3	27 ± 5	135
	3	9 ± 3	19 ± 8	211
	4	17 ± 5	39 ± 13	230
Thymus cells ($\times 10^7$)	1	73 ± 6	42 ± 5	58
	2	13 ± 3	10 ± 6	77
Thymus weight (mg)	1	25 ± 15	22 ± 15	88
	2	33 ± 12	26 ± 20	79
Spleen weight (mg)	1	98 ± 11	100 ± 9	102
	2	102 ± 9	78 ± 8	76

NOTE: For peripheral WBC, neutrocytes, and thymus cells, P value equals <0.05; for thymus weight and spleen weight, P value equals >0.05. Number of samples was three.
[a]Ratio of sarmentosin to control.
[b]White blood cells.

Table II. Inhibitory Effect of Sarmentosin on Number of Plaque-Forming Cells in Mice

Dosage (mg/kg × days)	*PFC[a] × 10⁴/Spleen*		
	Control	*Sarmentosin*	*S/C[b] × 100*
100 × 4	33 ± 9	42 ± 12	127
	56 ± 2	8 ± 2	14
	22 ± 18	23 ± 5	105
250 × 4	40 ± 9	35 ± 21	88
500 × 4	45 ± 4	19 ± 4	42
	31 ± 4	17 ± 10	55
	26 ± 6	12 ± 4	46

NOTE: P value for dosages of 100 and 250 equals >0.05, and P value for dosage of 500 equals <0.05. Number of samples was six.
[a]Plaque-forming cells.
[b]Ratio of sarmentosin to control.

Table III. Effect of Sarmentosin on Percentage of ANAF⁺ Cells in Various Tissues of Mice

Tissue	*Value*	*Expt. 1*	*Expt. 2*
Spleen	control	58 ± 3	54 ± 4
	sarmentosin	34 ± 8	60 ± 3
	S/C[a] × 100	59	11
	P value	>0.05	>0.05
Bone marrow	control	3 ± 1	4 ± 1
	sarmentosin	7 ± 3	11 ± 1
	S/C × 100	233	275
	P value	<0.01	<0.01
Peripheral WBC	control	49 ± 13	44 ± 1
	sarmentosin	40 ± 0	42 ± 5
	S/C × 100	82	95
	P value	>0.05	>0.05

NOTE: ANAF⁺ cells per 100 lymphocytes. Number of samples was three.
[a]Ratio of sarmentosin to control.

Table IV. Effect of Sarmentosin on Graft Versus Host Reaction of Rats

Dosage of Sarmentosin (mg/kg)	*Lymph Node Enlarged Index*			
	Control	*Sarmentosin*	*Inhibited (%)*	P *Value*
250	5.8 ± 2.8 (3)[a]	3.7 ± 1.4	36	>0.05
500	3.8 ± 1.8 (7)	1.0 ± 0.7	74	<0.05

[a]The number of animals is in parentheses.

In order to find the most active compound and to study the relationship between activity and structure, we recently began to synthesize a series of sarmentosin-type compounds. We have had some success with this ongoing project.

PSEUDOLARIC ACIDS A–C

Tu-jin-pi, the root barks of *Pseudolarix kaempferi* Gorden (Pinaceae), is traditionally used in Chinese medicine as an antifungal agent. The commercial product has the trade name Tujinpi tyinctura and is used in the treatment of *tinea pedis* (athlete's foot).

Three new antifungal components, pseudolaric acids A–C, were isolated from the barks. The root barks were first extracted with benzene, concentrated, and then extracted with a 5% solution of sodium bicarbonate. This extract was acidified with dilute mineral acid. A precipitate was obtained, collected, and chromatographed on a silica gel column with benzene and benzene–ethyl acetate. Pseudolaric acids A–C were readily isolated.

The following data were collected for pseudolaric acid B: $C_{23}H_{28}O_8$; mp 219 °C; $[\alpha]_D$ −46.4° (*c* 0.02, CH_3CH_2OH); HRMS (DI) *m/z* 414.1678 (M − H_2O), 372.1586 (M − CH_3COOH), 354.1588 (M − H_2O − CH_3COOH); CIMS *m/z* 433 (M + 1); UV 258 nm (log ϵ 4.5); IR 2500–3000, 1740, 1719, 1709, 1688, 1640 cm^{-1}. These data show the presence of an unsaturated acid, an ester, and a lactone ring, all of which are confirmed by four signals in the ^{13}C-NMR spectrum at δ 173.3, 172.8, 169.4, and 168.1. One of these signals is an acetyl group (δ_H 2.04). From the ^{13}C-NMR spectrum, double bond signals are observed at δ 144.5 d, 141.7 d, 138.7 d, 134.5 d, 127.9 s, and 121.8 d. These data and the molecular formula indicate that pseudolaric acid B should be a tricyclic diterpenic acid. Hydrolysis of the acid in aqueous potassium hydroxide solution afforded a diacid, $C_{22}H_{26}O_8$, and then a deacetyl diacid. Thus, acid B should contain one acid, one methyl ester, one acetyl, and one lactone ring. When the methyl ester of the deacetyl diacid was dehydrated with thionyl chloride and pyridine, a compound was obtained, and ozonolyses of pseudolaric acid B yielded pyruvic acid. When these results and ^1H-NMR double-resonance experiments are considered, acid B should consist of two partial structures.

The ^{13}C-NMR signals of pseudolaric acid B at δ 90.2 s and 83.7 s indicate two quaternary carbons bearing oxygens: one is carrying an acetoxyl group and another is carrying the lactone group. Pseudolaric acid B's methyl signal in the ^1H-NMR spectrum at δ 1.57 (3 H, s) reveals that the methyl is joined to the quaternary carbon bearing lactonic oxygen. These results lead to two possible structures for pseudolaric acid B. One structure could be excluded because of the lack of shield effect of Δ_{18} to CH_3-12.

Pseudolaric acid C [$C_{21}H_{26}O_7$; mp 208 °C; $[\alpha]_D$ −96.9° (*c* 0.024, CH_3OH)] is deacetylpseudolaric acid B. On methylation pseudolaric acid C afforded a dimethyl ester compound.

Pseudolaric acid A [$C_{22}H_{28}O_6$; mp 218-19 °C; $[\alpha]_D$ −46.4° (*c* 0.02, CH_3CH_2OH)], has very similar spectral data to that of pseudolaric acid B, but the proton signals are observed at δ 1.79 (3 H, CH_3C=C–) in pseudolaric acid A instead of 3.73 (3 H, s, $-COOCH_3$) as in pseudolaric acid B. The ^{13}C-NMR spectrum of pseudolaric acid A shows three carbonyl signals at δ 173.9 s, 173.7 s, and 170.4 s; six ethylenic carbon signals at δ 145.4 d, 139.5 s, 138.9 d, 129.9 s, 123.7 d, and 122.2 d; two signals of tertiary carbons bearing oxygen at δ 90.7 s and 83.2 s; four methyl signals at δ 28.0 q and 21.3 q; five methylene signals at δ 32.9 t, 30.3 t, 26.9 t, 26.4 t, and 24.0 t; one methyline signal at δ 48.9; and one quaternary carbon signal at δ 55.1 s. Pseudolaric acid A has UV absorption at 256 nm but does not display the shoulder near 230 nm. Therefore, the differences between pseudolaric acid B and pseudolaric acid A may be due to the methyl group in pseudolaric acid B instead of $-COOCH_3$ in pseudolaric acid A *(7–9)*. The proposed structures for pseudolaric acids A–C were confirmed by X-ray diffraction *(10)*.

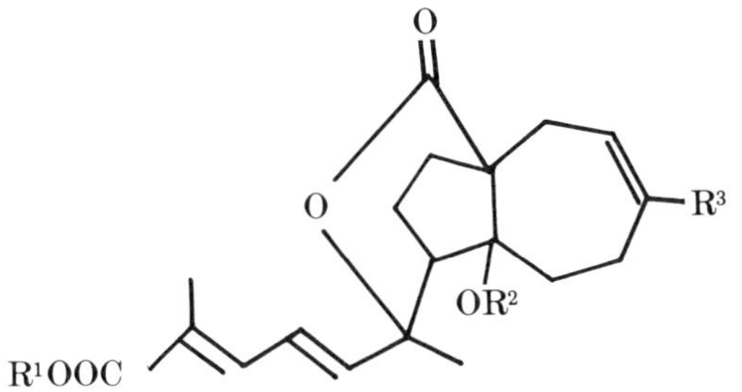

O O R^3 OR^2 R^1OOC

pseudolaric acid A, R^1 = H; R^2 = $COCH_3$; R^3 = CH_3
pseudolaric acid B, R^1 = H; R^2 = $COCH_3$; R^3 = $COOCH_3$
pseudolaric acid C, R^1 = H; R^2 = H; R^3 = $COOCH_3$

Mass spectroscopy studies on these acids and their derivatives led to the following observations and conclusions *(11)*.

1. All these compounds revealed very weak or nonexistent molecular ion peaks due to the great lability to loss of neutral fragments under electron impact.
2. Degradation patterns of these compounds were dependent on the substituents present at bridgehead position C-4. All compounds in the 4-$OCOOCH_3$ series were initiated by fission of the C-3–C-11 bond, leading to ions *m/z* 260 (a_2), 242 (a_4), 224 (a_5), 214 (a_6), 147 or 177 14*n* (b_1), and 131 (b_2). The preferential cleavage of ring B ($M^{+\cdot}$ → *m/z* 214 and

subsequent ions) was proven by incorporating *p*-Br-$C_6H_4COCH_2$ into the carboxyl terminal of the side chain as an indicative isotopic label. Completely different behaviors were observed in the 4-OH series (β-hydroxy lactones), where the initial reaction was triggered predominantly by C-3–C-4 bond cleavage followed by loss of H_2O, CO_2, R_2OH, or $HCOOR_2$ to give the ions C1–C4.

3. Both 4-hydroxy- or 4-acetoxy-substituted diterpenic acids gave m/z 110 + R_2 (d_1) and 111 (d_2) through cyclization-elimination.

In vitro the concentration of pseudolaric acid A for antifungal activity to *Oidium albicans* is about 200 ppm and that for pseudolaric acid B is 100 ppm *(12)*. Additional pharmacological tests either in (carboxymethyl)cellulose or in sodium bicarbonate solution showed pseudolaric acid B had a significant effect on terminating early pregnancy in rats, rabbits, and dogs at dosages of 10–40 mg/kg (Table V) *(13)*. However, implantation was not prevented in rats when pseudolaric acid B (40 mg/kg) was injected subcutaneously or intragastrointestinally (ig) daily 1–3 days after mating. Pseudolaric acid B also showed no estrogenic activity but caused severe desidual hemorrhage and necrosis. Pseudolaric acid B improved the plasma progesterone level after 5 days of administration at antifertility dose levels. Progesterone did not antagonize the effects of pseudolaric acid B on early pregnancy in rats.

To find a more active compound in the pseudolaric acid group, we synthesized a series of analogues (P-4–P-21, *see* page 186). Compounds P-4–P-16 were no more effective than pseudolaric acid B in terminating pregnancy in rats at a dosage of 15 mg/kg ig. The study is in progress *(14)*. Additional compounds are presently being prepared and tested.

GINNALINS A–C

Cha-tiao-qi, the leaves of *Acer ginnala* Maxim (Aceraceae), are used in folk medicine in Anhui province in the treatment of acute dysenteries *(15, 16)*. The leaves were extracted with 75% alcohol. After concentration of the extract, a crystalline precipitate, which we named ginnalin A, was separated. Ginnalin A was recrystallized from water, and the mother liquid was chromatographed on a

Table V. Effect of Pseudolaric Acid B on Early Pregnancy in Rats, Rabbits, and Dogs

	Medication			No. of Animals		No. of Fetuses ($X \pm SD$)	
Animals	*Route*	*Daily Dosage (mg/kg)*	*Days*	*Dosed*	*Pregnancy Terminated*	*Dead*	*Live*
Rats	sc	0	3	20	2	1 ± 1	7.9 ± 1.9
		25	3	10	10	4 ± 6	0
		30	3	25	22	6 ± 5	0.5 ± 1.6
		36	3	27	27	5 ± 4	0
	im	0	3	10	2	0	7.3 ± 4.1
		40	3	10	10	6 ± 5	0
	ig	0	3	20	4	0	7.1 ± 3.9
		15	3	10	9	6 ± 3	0.6 ± 1.8
		30	3	10	10	6 ± 3	0
Rabbits	sc	0	3	5	0	0	7 ± 2
		30	3	6	4	9 ± 4	1 ± 2
		40	3	4	4	4 ± 4	0
	iv	0	3	5	0	0	7 ± 2
		10	3	3	0	0	10 ± 3
		36	3	6	6	8 ± 5	0
	ig	0	3	5	0	0	7 ± 2
		10	3	7	0	2 ± 3	6 ± 3
		20	3	3	0	1 ± 2	7 ± 3
		40	3	5	2	2 ± 2	3 ± 3
Dogs	ig	0	3	2	0	0	5 ± 2
		5	3	5	5	4 ± 1	0
		20	1	4	4	necroses	0
		30	1	1	1	necroses	0

NOTE: Pseudolaric acid B was suspended in 1% (carboxymethyl)cellulose for ig or dissolved in 5% $NaHCO_3$ for sc, im, and iv.

silica gel column eluted with $CH_3CH_2OCOCH_3$:C_6H_6:CH_3CH_2OH:CH_3COOH (6:2:1:0.5), yielding ginnalins A–C. The isolated yield of A was 8.5%, B was 0.1%, and C was about 0.01%.

Ginnalin A, $C_{20}H_{20}O_{13}$, has the following physical data: mp 178–79 °C; $[\alpha]_D$ +30.9° [*c* 5.5, $(CH_3)_2SO$]; UV (log ϵ) 216 (4.7), 277 (4.42) nm; IR 3510, 3390, 3250, 1670, 1603, 1230, 1190, 1030 cm^{-1}. These data indicate the presence of a benzene ring, a conjugated ester, and hydroxyl groups. The UV spectrum of ginnalin A is very similar to that of gallic acid; however, the intensity is 1.3 times more intense. This fact suggests the presence of two galloyl residues in its molecule. The residues were verified by mineral acid hydrolysis, which gave 2 equiv of gallic acid and 1 equiv of

Pseudolaric Acid Analogues

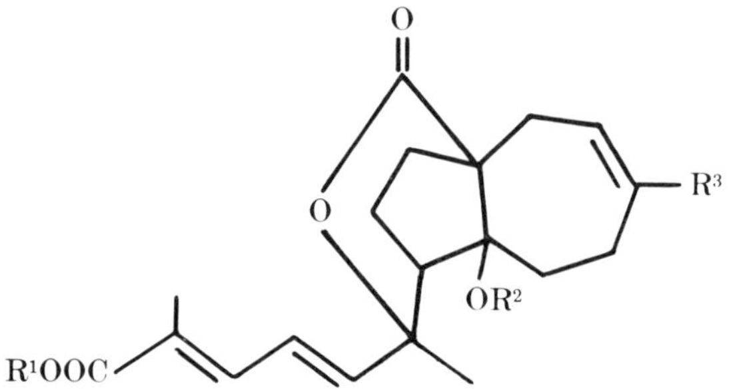

O
O
R^3
OR^2
R^1OOC

Compound	R^1	R^2	R^3
P-4	H	$COCH_3$	COOH
P-5	H	H	COOH
P-6	H	H	$COOCH_3$
P-7	H	H	$COOCH_2CH_3$
P-8	H	$COCH_3$	$COOCH_2CH_3$
P-9	H	H	$COOCH_2CH_2CH_3$
P-10	H	$COCH_3$	$COOCH_2CH_2CH_3$
P-11	H	H	$COOCH_2CH_2CH_2CH_3$
P-12	H	$COCH_3$	$COOCH_2CH_2CH_2CH_3$
P-13	$(CH_3)_2CH-$	$COCH_3$	$COOCH_3$
P-14	$(CH_3)_2CH-$	H	COOH
P-15	$(CH_3)_2CH-$	$COCH_3$	COOH
P-16	$(CH_3)_2CH-$	H	$COOCH_3$

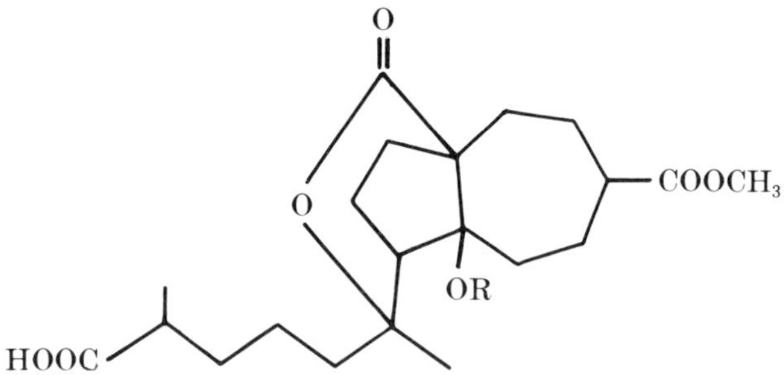

O
O
$COOCH_3$
OR
HOOC

P-17 R = $COCH_3$
P-18 R = H

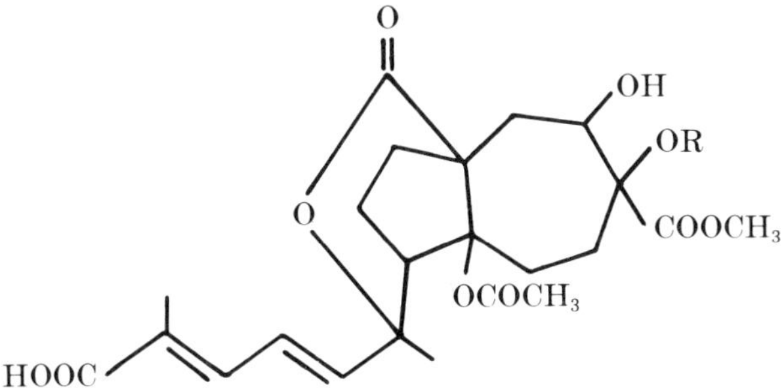

P-19 R = H

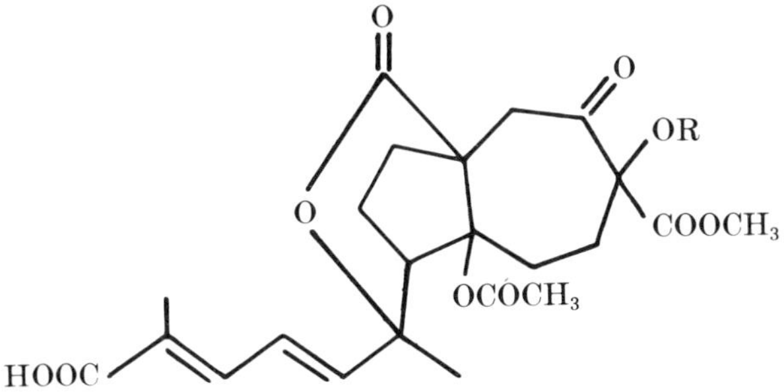

P-20 R = H
P-21 R = $COCH_3$

polygallitol. For determination of the position of the ester in the molecule, a hexamethyl ether solution of ginnalin A was oxidized with sodium periodate, reduced by sodium borohydride, and then hydrolyzed with mineral acid. Only glycerin was obtained. This fact indicates that two hydroxyl groups could be located at the C-3 and C-4 positions and thus two galloyl units joined to C-2 and C-6. Results of ^1H- and ^{13}C-NMR spectral analyses are consistent with the proposed structure for ginnalin A (Table VI) *(17)*.

Ginnalin B possesses the following physical data: $C_{13}H_{16}O_9$; mp 133–34 °C; $[\alpha]_D$ +14.8° [*c* 2.5, $(CH_3)_2CO$]; UV (log ϵ) 217 (4.36), 277 (3.94) nm. Acid hydrolysis of ginnalin B afforded 1 equiv each of gallic acid and polygallitol. As with ginnalin A, in its ^1H-NMR

Table VI. ^{1}H-NMR Data

Hydrogen	*Ginnalin A*	*Ginnalin B*
Ar-H	6.99 (2 H, s), 7.00 (2 H, s)	6.98 (2 H, s)
C-6-H	4.49 (1 H, d J = 12 Hz)	4.48 (1 H, dd, J = 11, 2 Hz)
C-6-H	4.24 (1 H, dd, J = 12, 5 Hz)	4.15 (1 H, dd, J = 11, 5 Hz)
C-2-H	4.76 (1 H, m)	3.88 (1 H, m)
C-1-He	3.95 (1 H, dd, J = 11, 5.1 Hz)	3.73 (1 H, dd, J = 10, 5 Hz)
C-1-Ha	3.27–3.60	3.03 (1 H, t, J = 10, 10 Hz)
C-3-H, C-4-H	3.27–3.60	3.16–3.42
C-5-H, H_2O	3.27–3.60	3.16–3.42
4′-Phenolic OH	8.92 (1 H, s), 8.96 (1 H, s)	—
3′,5′-Phenolic OH	9.28 (2 H, s), 9.32 (2 H, s)	—
Alcoholic OH	5.44 (2 H, s)	—

NOTE: Data in δ.

spectrum two H-6 signals appeared at low field, δ 4.30 and 4.48, due to the deshielding effect of galloyl. The signal of H-2 appeared at higher field, δ 3.88, due to lack of the shielding effect of galloyl.

Ginnalin C, mp 251–52 °C and $[\alpha]_D$ +64.8° [c 2.7, $(CH_3)_2SO$], has the same molecular formula and the same absorption as ginnalin B. When ginnalin C was hydrolyzed with mineral acid, 1 equiv each of gallic acid and polygallitol was afforded. Its H-2 signal in the ^{1}H-NMR spectrum is located at lower field (δ 4.65), indicating that the galloyl is located at C-2. Except for the H-6 signal, the other signals can be assigned as for ginnalin B.

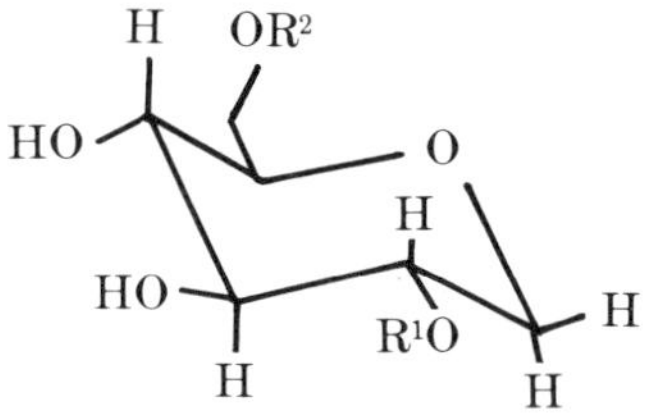

ginnalin A, $R^1 = R^2 =$

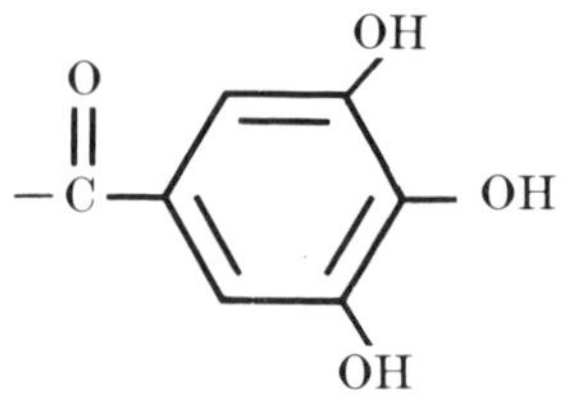

ginnalin B, R^1 = H; R^2 = —C(=O)—(3,4,5-trihydroxyphenyl)

ginnalin C, R^2 = H; R^1 = —C(=O)—(3,4,5-trihydroxyphenyl)

All these compounds and gallic acid, which was also separated from the leaves, show activity against *Shigella flexneri, Shigella sonnei,* and *Staphylococcus aureus* in vitro in concentrations of 0.20–0.50 mg/mL (Table VII). Ginnalin B is the most active compound. Ginnalin A was used in clinical trials in the treatment of 40 cases of acute dysentery. The rate of effect was 88% with a dosage of 2 g/day *(18)*.

Table VII. Antibacterial Activities of Ginnalins A and B In Vitro

Bacterium	*Ginnalin A*	*Ginnalin B*
Shigella flexneri	0.21	0.125
Shigella sonnei	—	0.25
Staphylococcus aureus	—	0.50
Pseudomonas aeruginosa	5.00	—

NOTE: Results are in milligrams per milliliter.

LITERATURE CITED

1. Fang, S. D.; Yan X. Q.; Li, J. F.; Fan, Z. Y.; Xu, R. S. *Kexue Tongbao* **1980,** *24,* 431.
2. Fang, S. D.; Yan X. Q.; Li, J. F.; Fan, Z. Y.; Xu, R. S. *Acta Chim. Sin.* **1982,** *40,* 273.
3. Zheng, C. D.; Qin, J. Z.; Gu, Y. X.; Zheng, Q. T. *Acta Phys. Sin.* **1981,** *30,* 242.
4. Gu, Y. X. *Acta Phys. Sin.* **1981,** *30,* 387.

5. Nahrstedt, A.; Wather, A.; Wray, V. *Phytochemistry* **1982,** *21,* 107.
6. Zhai, S. K.; Shen, M. L.; Xong, Y. L.; Li, J. F.; Ding, Yu. E.; Gao, Y. S. *Chin. J. Microbiol. Immunol.* **1982,** *2,* 145.
7. Cao, L. R.; Sun, Z. Y.; Wang, Z. D.; Yi, X. Z. *Zhonghua Pifuke Zazhi* **1957,** *5,* 286.
8. Li, Z. L.; Pan, D. J.; Hu, C. Q.; Wu, J. L.; Yang, S. S.; Xu, G. Q. *Acta Chim. Sin.* **1982,** *40,* 447.
9. Zhou, B. N.; Ying, B. P.; Song, Q. Q.; Chen, Z. X.; Han, J.; Yan, Y. F. *Planta Med.* **1983,** *47,* 35.
10. Yao, J. X.; Lin, X. Y. *Acta Chim. Sin.* **1982,** *40,* 385.
11. Hung, Z. H.; Liu, B. N.; Ying, B. P.; Yu, Q. T.; Han, J. *Acta Chim. Sin.* **1984,** *42,* 866.
12. Yang, G. L., unpublished results.
13. Wang, W. C.; Lu, R. F.; Zao, S. X.; Zhu, Y. Z. *Acta Pharmacol. Sin.* **1982,** *3,* 188.
14. Ying, B. P.; Han, J.; Xu, R. S., unpublished results.
15. Hong, S. H.; Song, C. Q.; Sheng, Y.; Zhang, F. J.; Xu, R. S. "Chemistry of Natural Products"; Science: Beijing, China, 1982; p. 244.
16. Song, C. Q.; Zhang, N.; Xu, R. S.; Song, G. Q.; Sheng, Y.; Hong, S. H. *Acta Chim. Sin.* **1982,** *40,* 1142.
17. Bock, K. *Phytochemistry* **1980,** *19,* 2033.
18. Song, C. Q., et al., unpublished results.

0939-1/86/0190$06.00/0

Chapter 13

Zingiberaceous Plants

PITTAYA TUNTIWACHWUTTIKUL*,
ORASA PANCHAROEN*,
DUNGTA KANJANAPOTHI†,
AMPAI PANTHONG†, WALTER C. TAYLOR‡,
and VICHAI REUTRAKUL§||

The Zingiberaceae is a large family of perennial herbaceous plants. The species are widely distributed throughout India, tropical Asia, and northern Australia. The family contains approximately 1400 species in 47 genera *(1)*. Larsen *(2)* has reported that approximately 200 species with some 29 genera occur in Thailand. However, this figure is rather uncertain because of the lack of sufficient information on the family due to the delicacy of the flowers, which are usually spoiled during the collection. In "Thai Plant Names," Smitinand *(3)* lists 78 species in 15 genera of the Zingiberaceae that are relatively common in Thailand (*see* Table I).

A number of the Zingiberaceous plants are used in Thai traditional medicine; for example, *Zingiber cassumunar* Roxb. is used as an antiinflammatory agent and is also used for relieving

*Department of Medical Science
Ministry of Public Health
Bangkok 10100, Thailand

†Department of Pharmacology
Faculty of Medicine
Chiangmai, Thailand

‡Department of Organic Chemistry
University of Sydney
Sydney, New South Wales, 2006, Australia

§Department of Chemistry
Faculty of Science, Mahidol University
Bangkok 10400, Thailand

||Author to whom correspondence should be addressed

asthmatic symptoms, *Alpinia galanga* Sw. and *Curcuma longa* Linn. are used externally for the treatment of skin diseases, and *Curcuma xanthorrhiza* Roxb. is an emmenagogue. Many of the plants of this family are used very commonly as flavoring agents, such as *Amomum krervanh* Pierre, *Alpinia siamensis* Schum., *Alpinia galanga* Sw., and *Zingiber officinale.*

The importance of Zingiberaceous plants in traditional medicine throughout the tropical world is apparent from the computer printout of the NAPRALERT data base *(4)*. Not less than 56 different applications to medical problems are listed. Many of the

Table I. Genera of the Zingiberaceae Commonly Found in Thailand

Genus	*Number of Species*
Alpinia	5
Amomum	7
Boesenbergia	4
Catimbium	2
Caulokaempferia	6
Cenolophon	1
Costus	2
Curcuma	14
Elettaria	1
Globba	9
Hedychium	8
Kaempferia	7
Languas	1
Nicolaia	3
Zingiber	9

usages are quite possibly without any medical basis or value, but some are sufficiently consistent and widespread as to demand further scientific study. The biological activities of certain species of the Zingiberaceae in different animal test systems have been studied and reported. The box lists some of the important ethnomedical applications *(4)*, and Table II summarizes some of the significant biological-activities testing performed on the 10 Zingiberaceous plant species that appear to have been examined most closely.

Selected Ethnomedical Applications of Zingiberaceous Plants

Abortifacient	Aphrodisiac
Antiinflammatory	Antitussive
Antimalarial	Cardiotonic
Antitumor	Cure for gastrointestinal disorders
Antipyretic	Insecticide
Antifungal	

Table II. Biological Activities of Selected Zingiberaceous Species

Species	*Type of Biological Activity*	*Ref.*
Alpinia galanga	antibacterial, smooth muscle stimulant	4, 5
Costus speciosus	uterine stimulant	4, 6
Curcuma aromatica	abortifacient, antifungal	4
Elettaria cardamomum	cytotoxic, antispasmodic	4
Hedychium spicatum	antibacterial	4, 7
Kaempferia galanga	anthelmintic, mutagenic	4
Zingiber capitatum	cytotoxic	4, 8
Zingiber cassumunar	smooth muscle relaxant, antiinflammatory	4, 9, 10
Zingiber officinale	inhibitors of prostaglandin biosynthesis	4, 11
Zingiber zerumbet	cytotoxic	4, 12

The Zingiberaceous plants have been proven to be a rich source of substances of phytochemical interest. In addition to substances belonging to the common classes of secondary natural products (terpenes, flavonoids, etc.), a sizable number of substances having novel structures have been isolated recently, and this situation is stimulating chemical interest in the family. In a number of instances, pharmacological and biological evaluations have been made on substances isolated by normal procedures of extraction and fractionation, while occasionally a biological assay has been used to follow fractionation of extracts in order to isolate biologically active materials. *Alpinia* sp. has been extensively investigated from both the phytochemical and biological points of view. The seeds of *Alpinia japonica* have been used as an aromatic stomachic in Japan. From the fresh rhizome of the plant, a novel sesquiterpene peroxide was isolated together with 3α,4α-oxidoagarofuran, furopelarogan B, 4α-hydroxydihydroagarofuran, α-agarofuran, 10-*epi*-γ-eudesmol, and β-eudesmol *(13)*. Sesquiterpere peroxide undergoes rearrangement and cyclization to give furopelarogan B.

Diarylheptanoids were isolated from the rhizomes of *Alpinia officinarum (14, 15)*. Several diarylheptanoids were shown to be inhibitors of prostaglandin biosynthesis.

In addition to these compounds, the essential oil from which 78 components were identified has been isolated *(16)*. Two diarylheptanoids have been isolated from *Alpinia oxyphylla (17, 18)*. One compound was reported to be 125 times more pungent than zingerone and also was one of the most potent inhibitors of prostaglandin biosynthesis. From the fruits of *A. oxyphylla,* the monoterpenes pinene, camphor, and 1,8-cineol; and sesquiterpenes zingiberene and zingeriberol were isolated.

The seeds of *Alpinia katsumadai* have been used as an aromatic stomachic in Japan. The constituents of the seeds are 1,8-cineole; α-humulene; *trans,trans*-farnesol; linalool; camphor; terpinen-4-ol; carvotanacetone; bornyl acetate; geranyl acetate; methyl cinnamate; nerolidol; and three known flavonoids, alpinetin, pinocembrin, and cardamonin *(19)*. On further investigation of the seeds, six diarylheptanoids have been isolated *(20)*.

The methanolic extract of the rhizomes of *Alpinia speciosa* possesses potent antihistaminic activity. Early work on the rhizomes of the plant led to the isolation of dihydro-5,6-dehydrokawain, *(21, 22)* and alpinetin and cardamonin *(23)*. Recently, Itokawa *(24)* demonstrated that the methanolic extracts of the rhizomes of *Alpinia speciosa* possess significant inhibitory activities against histamine and barium chloride by the Magnus method using excised guinea pig ileum. This observation led to the isolation of two diterpenes, methyl *trans*-cinnamate, alpinetin, cardamonin, dihydro-5,6-dehydrokawain, and three other compounds *(24)*.

From the fresh root of *Alpinia flabellata,* an interesting compound, alflabene, was isolated *(25)*. The same compound was also found in the rhizome of *Zingiber cassumunar.*

Methanolic extracts of seeds of *Alpinia galanga (26)* showed significant inhibitory activity against shay ulcer in rats. Bioassay-guided fractionation revealed that the inhibitory activity was concentrated in the fraction containing ether-soluble neutral substances. Further fractionation involving silica gel chromatography yielded the two antiulcer substances 1′-acetoxychavicol acetate and 1′-acetoxyeugenol acetate; together with sesquiterpenes caryophyllene oxide, caryophyllenol I, and caryophyllenol II.

In addition to the *Alpinia* species mentioned, 10 more species have also been recorded. *A. caerulea, A. elatior, A. formosana, A. gracilis, A. malaccensis, A. melanocarpa, A. mutica, A. nutans, A. purpurata,* and *A. romburghiana* have been analyzed for flavonoids *(27)*. Very little work has been carried out on other species, that is, *A. calcaratta, A. chinensis, A. khulanjan, A. kumatake, A. roxburghii,* and *A. uviformis (4)*.

The roots of *Afromomum daniellii* are used in Africa as a purgative. Chemical investigation led to the isolation of the diterpene dialdehyde (*E*)-8β,17-epoxylabd-12-ene-15,16-dial *(28)*. *Afromomum giganteum* is rich in terpenes *(29)* and phenolic compounds *(30, 31)*. The composition of the essential oil of the stems of *Af. giganteum* was investigated; 47 terpenes were identified *(29)*. Three rare naturally occurring flavonol 3-methyl ethers, kaemferol with the 3,7,4′-hydroxys replaced with 3,7,4′-methyls; quercetin with the 3,7,4′-hydroxys replaced with 3,7,4′-methyls; and quercetin with the 3,7,3′,4′-hydroxys replaced with 3,7,3′,4′-methyls have been isolated together with chrysophanol, physcion, 2,6-dimethoxybenzoquinone, and β-sitosterol from the stems of this plant. *(30)*. Further investigation of the chloroform extract of the stems of the plant has yielded emodin, syringaldehyde, syringic acid, and dehydrozingerone *(31)*. Investigation of the essential oil of the seeds of *Afromomum mala* has revealed that the constituents are predominantly monoterpenoids α-pinene, β-pinene, sabinene, α-phellandrene, limonone, and 1,8-cineole *(32)*. The constituents of the seeds of *Afromomum pruinosum* were identified to be flavonoids *(33)*. Other *Afromomum* species, that is, *Af. angustifolium, Af. baumannii, Af. korarima, Af. sanguineum,* and *Af. simiarum,* are also recorded *(4)*; however, their chemical constituents and biological activities are not well known.

The essential oil of the fruits of *Amomum cardamomum* contains 15 mono- and sesquiterpenes, and the oil of *Amomum globosum* is a mixture of 23 mono- and sesquiterpenes *(34)*. Two indane aldehydes along with seven terpenoids have been isolated from the ether extract and the oil of the seeds of *Amomum medium (35)*. Chemical investigation of the acetone extract of the seeds of *Amomum melegueta* reveals the presence of hydroxyphenylalkanones *(36)*. In an earlier study on the same plant, two gingerol derivatives were also identified *(37)*. The oil of the dry fruits of *Amomum subultatum* contains 13 monoterpenes *(38)*. An aurone glycoside *(39, 40)*, cardamonin *(41)*, and alpinetin *(41)* have been isolated from the seeds of *Am. subulatum.* Investigations on chemical constituents and biological activities of several *Amomum* sp. are not well known. These species are *Am. aromaticum, Am. aurantiacum, Am. costatum, Am. dallachyi, Am. dealbatum, Am. gracile, Am. hochreutineri, Am. krervahn, Am. korarima, Am. longiligare, Am. repens, Am. villosum,* and *Am. xanthioides (4)*.

The chemistry of the genus *Costus* appears to be somewhat distinctive from other genera of the Zingiberaceae. The *Costus* species are found to be rich in steroidal saponins and sapogenins. The rhizomes of *Costus speciosus* have been reported to contain more than 2% of diosgenin on the basis of the dry rhizomes *(6, 42)*. Diosgenin is one of the sapogenins extensively used in the manufacture of steroidal hormones. The occurrence of diosgenin in the rhizomes of *C. speciosus* has attracted a great deal of further investigations of the plants *(43–47)*. Recent works have reported the isolation of aliphatic hydroxy ketones and a new sterol *(48, 49)*, and two new benzoquinones together with α-tocopherolquinone from its rhizome *(50)*. The fresh fruit juice of *Costus afar* has been reported to exhibit antiinflammatory activity *(51)*. Chemical investigation of the rhizomes of the plant reveals the presence of various sapogenins, namely, diosgenin, stigmasterol, and costugenin *(52)*.

A number of species of the genus *Curcuma* contain a mixture of monoterpenes, sesquiterpenes, and C_6–C_3 compounds; however, some are rich in novel sesquiterpenes and others are rich in biarylheptanoids. *Curcuma aromatica* and *Curcuma longa* are reported to contain curcumin *(53, 54)*. Curcumin possesses antiinflammatory activity, and clinical trial of the substance has been carried out *(55)*. Studies on the biosynthesis of curcumin have been reported *(56, 57)*. *Cur. aromatica* possesses a wide range of activities, that is, abortifacient, antifungal, antispasmodic, and antitumor *(4)*. Several monoterpenes and sesquiterpenes, including α-curcumene, are isolated from the species *(58)*. *Cur. longa* has been used as an aromatic stomachic, as a diuretic, and as a remedy for jaundice. Recently, several curcuminoids isolated from *Cur. longa* rhizomes demonstrated significant antihepatotoxic activity *(59)*. The essential oil obtained from the rhizomes of *Curcuma wenyujin* is found to have antitumor activity. Curcumol is one of the constituents of this rhizome; its absolute stereostructure has been determined *(60)*. *Curcuma zedoaria* has been used as an abortifacient and antifungal agent. In China the rhizomes of *Curcuma zedoaria* are used clinically in the treatment of several types of tumors. Chemical investigations of the rhizomes reveal the presence of several sesquiterpenoids, namely, curcumol *(61)*, zederone *(62)*, curdione *(63)*, curcolone *(64)*, curcumenol *(65)*, furanodiene, curzerene *(66)*, curzerenone, epicurzerenone *(67)*, procurcumenol *(68)*, isocurcumenol *(69)*, and curcumadiol *(70)*. A sesquiterpene, xanthorrhizol, has been isolated from the rhizomes of *Curcuma zanthorrhiza (71)*.

Zingiber officinale is grown commercially in most tropical regions for its rhizome, ginger, which is valued for its flavor and pungency. Ginger is used in medicine as a carminative and stimulant for the gastrointestinal tract. A very extensive bibliography covering the work done on the plant (up to 1981) has been published *(72)*. The biological activity of ginger is the most extensively studied among the Zingiberaceae. The extract of *Z. officinale* has been tested for a number of different biological activities, that is, larvacidal, antihypercholesterolemic, narcotic antagonist, cytotoxic, antibacterial, anticonvulsant, analgesic, antiulcer, gastric antisecretor, antitumor, molluscicidal, antifungal, antispasmodic, and antiallergenic *(4)*. Recently, the powdered rhizome of *Z. officinale* was demonstrated to be superior to dimenhydrinate in preventing the gastrointestinal symptoms in motion sickness *(73)*. The species is a rich source of terpenoids. The essential oil of the rhizome contains at least 22 terpenes *(74, 75)*. Chemical investigations of the rhizome also reveal the presence of various sesquiterpenoids, (-)-zingiberene *(76)*, (-)-β-sesquiphellandrene *(77)*, sesquithujene, *cis*-sesquisabinene hydrate, and zingiberenol *(78)*. Much attention has been paid to the pungent principles of ginger, which are known to be arylalkanol derivatives. The existence of [3]- to [8]-, [10]-, and [12]-gingerols in the rhizome of *Z. officinale* has been demonstrated *(79–82)*. In addition to the gingerols, other constituents are also isolated from the roots of *Z. officinale.* They are identified as [6]- and [10]-dehydrogingerdiones and [6]- and [10]-gingerdiones and are found to be potent inhibitors of prostaglandin biosynthesis *(83)*. Syntheses of the pungent principles of ginger and related compounds have been carried out *(84, 85)*. Studies in the biosynthesis of [6]-gingerol, a pungent constituent of ginger, have been reported *(86, 87)*.

The isolation of compounds [6]-dehydrogingerdione and [6]-gingerdione along with compounds [10]-dehydrogingerdione and [10]-gingerdione from *Z. officinale* gives strong support to the biosynthetic pathway for [6]-gingerol as proposed by Denniff et al. *(86)*. The essential oil of the rhizomes of *Zingiber zerumbet* contains, in addition to monoterpenes *(87)*, sesquiterpenes including β-caryophyllene, *ar*-curcumene, zerumbone, humulene, humulene monoxide, and humulene dioxide *(88–90)*. Several cytotoxic substances, zerumbone, zerumbone epoxide, three curcuminoids, and 3,4-diacetylafzelin are found in the rhizomes of *Z. zerumbet* *(91)*.

Not a great deal of work has been reported on the genera *Elettaria, Globba, Hedychium* and *Kaempferia.* However, cryptomeridiol *(92)*, hedychenone *(93)*, and 7-hydroxyhedychenone *(94)* have been isolated from the rhizomes of *Hedychium spicatum;* the oxygenated cyclohexane derivative, crotepoxide, has been isolated from the roots of *Kaempferia rotunda (95)*; and at least 29 terpenoids have been isolated from the oil of the seeds of *Elettaria cardamomum (96)*.

The survey shows that the plants in the Zingiberaceae are capable of producing a wide range of interesting secondary metabolites. In reality, only a comparatively small proportion of the species has been chemically examined, and a clear chemotaxonomic picture of the family at the genera level, as well as at the species level, has not yet emerged. For certain, much novel chemistry awaits to be uncovered.

Today, the family is characterized by the production of the following chemical constituents: (1) terpenoids, including monoterpenoids, sesquiterpenoids, and diterpenoids; (2) flavonoids; (3) diarylheptanoids, which occur in only one other family (Butulaceae) *(97)*; and (4) arylalkanones, which are found only in the Zingiberaceae.

Our research group has been interested in plants of Zingiberaceae due to their many uses in Thai traditional medicine. On the basis of their alleged pharmacological activities, several species of Zingiberaceous plants, namely, *Boesenbergia pandurata* (four different varieties: yellow, red, black, and white rhizome) and *Zingiber cassumunar,* have been chosen for investigation in detail in our laboratory.

B. pandurata (red rhizome) has been used in folk medicine for the treatment of colic. The ethanol extract of the milled rhizomes possesses intestinal smooth muscle relaxation activity. Chemical investigation of the hexane and chloroform extracts of the rhizomes yielded (±)-pinostrobin, (±)-pinocembrin, boesenbergin A, 2′,6′-dihydroxy-4′-methoxychalcone, rubranine, panduratin A, and a pair of diastereomers, panduration B_1 and panduration B_2 *(98, 99)*. An improved synthesis of rubranine *(100)*, a substance previously isolated from *Aniba rosaeodora (101)*, from pinocembrin and citral was developed. Panduratin A, a cyclohexenyl derivative, is, so far, unique to this family. The same type of cyclohexenyl moiety has been reported to occur only in one other family (Moraceae) *(102)*. Pinocembrin has been shown to exhibit smooth muscle relaxation activity.

The hexane extract of the black rhizomes of *Boesenbergia pandurata* exhibited potent smooth muscle relaxation activity and antiinflammatory action in the carrageenin-induced pedal edema test *(103)*. Eleven flavonoids have been isolated from the hexane extract of the rhizomes, and only one compound, (±)-pinostrobin, is common to both the black and the red rhizome varity of *B. pandurata (104)*.

5,7-Dimethoxyflavanone shows strong antiinflammatory activity in the carrageenin-induced pedal edema test.

The white rhizome of *B. pandurata* has also been chemically investigated in our laboratory. Examination of the chloroform extract of the rhizomes has yielded two oxygenated cyclohexane derivatives, crotepoxide and boesenboxide, as well as isopimaric acid and 2′-hydroxy-4,6′,4′-trimethoxychalcone *(99, 105)*. The structure of crotepoxide has been confirmed by single-crystal X-ray analysis *(106)*.

B. pandurata Schl. (yellow rhizome) is commonly known in Thailand as krachai. The rhizome is used as a flavoring agent in Thai cooking and also in folk medicine for the treatment of colic and as an aphrodisiac. The ethanolic extract of the dry rhizome possesses smooth muscle relaxant activity.

The essential oil obtained contains mainly monoterpenes *(107)*. Mongkolsuk and Dean *(108)* have isolated (±)-pinostrobin and (±)-alpinetin from the dry rhizomes. Recent investigation of the chloroform extract of the fresh rhizomes yielded (±)-pinostrobin; (±)-pinocembrin; two known chalcones, 2′,6′-dihydroxy-4′-methoxychalcone and cardamonin; boesenbergin A; boesenbergin B; panduratin A; and a pair of diastereomers, panduratins B_1 and B_2.

Z. cassumunar has long been used in Thai traditional medicine as an agent for relieving muscle pain and inflammation. The pulverized rhizomes have been given orally for relieving asthmatic symptoms. Biological testing of the alcoholic extract of the rhizome showed antispasmodic activity on smooth muscle *(9, 109)*. Solvent partition of the extract located the activity in the hexane fraction. The hexane, chloroform, and methanol extracts of the rhizomes were subjected to an antiinflammatory test using the carrageenin-induced rat paw edema method, and the activity was located in the hexane extract *(10)*.

A recent study of the antiasthmatic activity of *Z. cassumunar* in asthmatic children was conducted. Orally administered capsules containing the pulverized rhizome of *Z. cassumunar* showed a positive effect on relieving asthmatic symptoms and low side effects.

The essential oil of the rhizome of *Z. cassumunar* was found to contain α-pinene, β-pinene, sabinene, myrcene, α-terpinene, limonene, γ-terpinene, ρ-cymene, terpinolone, and terpinen-4-ol *(111)*.

Chemical investigation of the hexane extract of the milled rhizomes yielded 13 aromatic compounds *(112–14)*. Structures of these compounds were established on the basis of their spectroscopic data and syntheses. The structure and stereochemistry confirmation of one compound was achieved by single-crystal X-ray diffraction analysis. Another compound exhibits anti-inflammatory activity in the carrageenin-induced pedal edema test.

After our results were published, two other groups [German *(115)* and Japanese *(116)*] also published their chemical investigations of *Z. cassumunar,* which confirmed our findings.

The occurrence of several compounds is noteworthy. These compounds have a C_6–C_4-type skeleton, which is rare in nature. Only few species were reported to contain this type of carbon skeleton: These species are *Aframomum giganteum* (dehydrozingerone), *Artemisia campestris* subsp. *glutinosa (117)* [4-(*o*-hydroxyphenyl)butan-2-one], and *Z. officinale (118)* (zingerone). These compounds are very interesting from the biosynthetic aspect. The study of the biosynthetic origin of these compounds would be highly desirable.

Both the chemistry and pharmacology of zingiberaceous plants have proved to be highly interesting. No doubt much information on the plants of this family is yet to be discovered.

LITERATURE CITED

1. Holtum, R. E. *Gard. Bull. (Singapore)* **1950,** *13,* 1.
2. Larsen, K. *Nat. Hist. Bull. Siam. Soc.* **1980,** *28,* 151.
3. Smitinand, T. "Thai Plant Names: Botanical Names—Vernacular Names"; Royal Forest Department: Thailand, 1980.
4. NAPRALERT print-out, "Ethnomedical, Pharmacological and Phytochemical Survey on Plants of the Zingiberaceae," The University of Illinois at Chicago, Program for Collaborative Research in the Pharmaceutical Sciences, 1984. We are grateful to Professor N. R. Farnsworth for providing this information from NAPRALERT data base.
5. Mokklasmit, M.; Ngwarmwanthana, W.; Swasdimongkol, K.; Permphiphat, U. *J. Med. Assoc. Thailand* **1971,** *54,* 490–503.
6. Dasgupta, B.; Pandey, V. B. *Experientia* **1970,** *26,* 475.
7. Ray, P. G.; Mujumder, S. K. *Econ. Bot.* **1976,** 317.
8. Dhawan, B. N.; Patnaik, G. K.; Rastogi, R. P.; Singh, K. K.; Tandon, J. S. *Indian J. Exp. Biol* **1977,** *15,* 208.
9. Anantasan, V.; Mopadonratanakhun, L. *J. Natl. Res. Counc. Thailand* **1980,** *12,* 51.
10. Malone, M., Univ. of the Pacific, Stockton, Calif., private communication.
11. Kiuchi, F.; Shibuya, M.; Sankawa, U. *Chem. Pharm. Bull.* **1982,** *30,* 754–57.
12. Mattes, H. W. D.; Luu, B.; Ourisson, G. *Phytochemistry* **1980,** *19,* 2643.
13. Itokawa, H.; Watanabe, K.; Mihashi, S.; Iitaka, Y. *Chem. Pharm. Bull* **1981,** *28,* 681.
14. Itokawa, H.; Morita, M.; Mihashi, S. *Chem. Pharm. Bull.* **1981,** *29,* 2383.
15. Kiuchi, F.; Shibuya, M.; Sankawa, U. *Chem. Pharm. Bull.* **1982,** *30,* 2279.
16. Lawrence, B. M.; Hogg, J. W. *Perfum. Essent. Oil Rec.* **1969,** *60,* 88.
17. Itokawa, H.; Aiyama, R.; Ikuta, A. *Phytochemistry* **1981,** *20,* 769.
18. Itokawa, H.; Aiyama, R.; Ikuta, A. *Phytochemistry* **1982,** *21,* 241.
19. Saiki, Y.; Ishikawa, Y.; Uchida, M.; Fukushima, S. *Phytochemistry* **1978,** *17,* 808.
20. Kuroyanagi, M.; Koro, T.; Fukushima, S.; Aiyama, R.; Ikuta, A.; Itokawa, H.; Morita, M. *Chem. Pharm. Bull.* **1983,** *31,* 1544.

21. Kimura, Y.; Takahashi, S.; Yoshida, I. *Yakugaku Zasshi* **1968,** *88,* 239; *C.A.* **1968,** *69,* 35863k.
22. Kimura, Y.; Takido, M.; Nakano, K.; Takishita, M. *Yakugaku Zasshi* **1966,** *86,* 1184; *C.A.* **1967,** *67,* 21775e.
23. Krishna, B. M.; Chaganty, R. B. *Phytochemistry* **1973,** *12,* 238.
24. Itokawa, H.; Morita, M.; Mihashi, S. *Chem. Pharm. Bull.* **1980,** *28,* 3452; *Phytochemistry* **1981,** *20,* 2503.
25. Mori, I.; Nakachi, Y.; Ueda, K.; Uemura, D; Hirata, Y. *Tetrahedron Lett.* **1978,** 2297.
26. Mitsui, S.; Kobayashi, S.; Nagahori, H.; Ogiso, A. *Chem. Pharm. Bull.* **1976,** *24,* 2377.
27. Williams, C. A.; Harborne, J. B. *Biochem. Syst. Ecol.* **1977,** *5,* 221.
28. Kimbu, S. F.; Nijmi, T. K.; Sordangam, B. L.; Akinniyi, J. A.; Connolly, J. D. *J. Chem. Soc. Perkin Trans. 1* **1979,** 1303.
29. de Bernardi, M.; Mellerio, G.; Colombo, M. P.; Vidari, G.; Vita-Finzi, P. *Planta Med.* **1981,** *41,* 359.
30. de Bernardi, M.; Vidari, G.; Vita-Finzi, P. *Phytochemistry* **1976,** *15,* 1785.
31. Vidari, G.; Vita-Finzi, P.; de Bernardi, M. *Phytochemistry* **1971,** *10,* 3335.
32. Eglinton, G.; Hamilton, R. J. *Phytochemistry* **1965,** *4,* 197.
33. Ayafor, J. F.; Connolly, J. D. *J. Chem. Soc. Perkin Trans. 1* **1981,** 2563.
34. Lawrence, B. M.; Hogg, J. W.; Terhune, S. J. *Phytochemistry* **1972,** *11,* 1534.
35. Takido, M.; Yoshikawa, Y.; Yamanouchi, S.; Kimura, Y. *Phytochemistry* **1978,** *17,* 327.
36. Tackie, A. N.; Dwuma-Badu, D.; Agim, J. S. K.; Dabra, T. T.; Knapp, J. E.; Slatkin, D. J.; Schiff, P. L., Jr. *Phytochemistry* **1975,** *14,* 853.
37. Connell, D. W. *Aust. J. Chem.* **1970,** *23,* 369.
38. Lawrence, B. M. *Phytochemistry* **1970,** *9,* 665.
39. Lakshmi, V.; Chauhan, J. S. *J. Indian Chem. Soc.* **1976,** *53,* 633.
40. Lakshmi, V.; Chauhan, J. S.; *Indian J. Chem. Sect. B* **1977,** *15,* 814.
41. Bheemasankara, C.; Namosiva, T.; Suryaprakasam, S. *Planta Med.* **1976,** *29,* 391.
42. Sarin, Y. K.; Kapahi, B. K.; Kapur, S. K.; Atal, C. K. *Curr. Sci.* **1976,** *42,* 688.
43. Sarin, Y. K.; Bedi, K. L.; Atal, C. K. *Curr. Sci.* **1974,** *43,* 569.
44. Tschesche, R.; Pandey, V. B. *Phytochemistry* **1978,** *17,* 1781.
45. Shah, C. S.; Bhavsar, G. C.; Seth, A. *Curr. Sci* **1978,** *47,* 270.
46. Rathore, A. K.; Khannce, P. *J. Nat. Prod.* **1978,** 640.
47. Gupta, M. M.; Farocqui, S. V.; Lal, R. N. *J. Nat. Prod.* **1981,** *44,* 486.
48. Gupta, M. M.; Lal, R. N.; Shukla, Y. N. *Phytochemistry* **1981,** *20,* 2553, 2557.
49. Gupta, M. M.; Lal, R. N.; Shukla, Y. N. *Phytochemistry* **1982,** *21,* 230.
50. Mamhood, U.; Shukla, Y. N.; Thakur, R. S. *Phytochemistry* **1984,** *23,* 1725.
51. Iwu, M. M.; Anyannu, B. N. *J. Ethnopharmacol.* **1982,** *6,* 263.
52. Iwu, M. M. *Planta Med.* **1981,** *43,* 413.
53. Rao, B. S.; Shintre, V. P. *J. Soc. Chem. Ind.* **1928,** *47,* 541; *C.A.* **1928,** *22,* 1826.
54. Janaki, N.; Bose, J. L. *J. Indian Chem. Soc.* **1967,** *44,* 985.
55. de Souza, N. J.; Ganguli, B. N.; Reden, J. *Annu. Rep. Med. Chem.* **1982,** *17,* 301.
56. Roughley, P. J.; Whiting, D. A. *Tetrahedron Lett.* **1971,** 3741.
57. Roughley, P. J.; Whiting, D. A. *J. Chem. Soc. Perkin Trans 1* **1973,** 2379.
58. Honwad, V. K.; Rao, A. S. *Tetrahedron* **1965,** *21,* 2593.
59. Kiso, Y.; Suzuki, Y.; Watanabe, V.; Oshima, Y.; Hikino, H. *Planta Med.* **1983,** *49,* 185.
60. Inayama, S.; Gao, J. F.; Harima, K.; Kawamata, T.; Iitaka, Y.; Guo, Y. T. *Chem. Pharm. Bull.* **1984,** *32,* 3783.

61. Hikino, H.; Meguro, K.; Sakurai, Y.; Takemoto, T. *Chem. Pharm. Bull.* **1966,** *14,* 1241.
62. Hikino, H.; Torik, K.; Horibe, I.; Kuriyama, K. *J. Chem. Soc. C* **1971,** 688.
63. Hikino, H.; Sakurai, Y.; Takahashi, H.; Takemoto, T. *Chem. Pharm. Bull.* **1967,** *15,* 1390.
64. Hikino, H.; Sakurai, Y.; Takemoto, T. *Chem. Pharm. Bull.* **1968,** *16,* 827.
65. Hikino, H.; Meguro, K.; Sakurai, Y.; Takemoto, T. *Chem. Pharm. Bull.* **1965,** *13,* 1484; **1966,** *14,* 1241.
66. Hikino, H.; Agatsuma, K.; Konno, C.; Takemoto, T. *Chem. Pharm. Bull.* **1970,** *18,* 752.
67. Hikino, H.; Agatsuma, K.; Takemoto, T. *Tetrahedron Lett.* **1968,** 2855.
68. Hikino, H.; Sakurai, Y.; Takemoto, T. *Chem. Pharm. Bull.* **1968,** *16,* 1605.
69. Hikino, H.; Agatsuma, K.; Takemoto, T. *Chem. Pharm. Bull.* **1969,** *17,* 959.
70. Hikino, H.; Konno, C.; Takemoto, T. *Chem. Pharm. Bull.* **1971,** *19,* 93.
71. Rimpler, H.; Haensel, R.; Kochendoerfer, L. *Z. Naturforschung B* **1970,** *25,* 995; *C.A.* **1971,** *74,* 13288C.
72. Bibliography *Med. Aromat. Plants Abstr.* **1981,** *3,* 65–79.
73. Mowrey, D. B.; Clayson, D. E. *Lancet* **1982,** 655.
74. Kami, T.; Makayama, M.; Hayashi, S. *Phytochemistry* **1972,** *11,* 3377.
75. Smith, R. M.; Robinson, J. M. *Phytochemistry* **1981,** *20,* 203.
76. Eschenmoser, A.; Shinz, H. *Helv. Chim. Acta* **1950,** *33,* 171.
77. Cannell, D. W.; Sutherland, M.D. *Aust. J. Chem.* **1966,** *19,* 283.
78. Terhune, S. J.; Hogg, J. W.; Bronstein, A. C.; Lawrence, B. M. *Can. J. Chem.* **1975,** *53,* 3285.
79. Connell, D. W.; Sutherland, M. D. *Aust. J. Chem.* **1969,** *22,* 1033.
80. Masada, Y.; Inoue, T.; Hashimoto, K.; Fujioka, M.; Shiraki, K. *J. Pharm. Soc. Jpn.* **1973,** *93,* 318.
81. Masada, Y.; Inoue, T.; Hashimoto, K.; Fujioka, M.; Uchino, C. *J. Pharm. Soc. J.* **1974,** *94,* 734.
82. Murata, T.; Shinohara, M.; Miyamoto, M. *Chem. Pharm. Bull.* **1972,** *20,* 2291.
83. Kiuchi, F.; Shibuya, M.; Sankawa, U. *Chem. Pharm. Bull.* **1982,** *30,* 754.
84. Denniff, P.; Macleod, I.; Whiting, D. A. *J. Chem. Soc. Perkin Trans. 1* **1981,** 82.
85. Kato, N.; Hamada, Y.; Shioiri, T. *Chem. Pharm. Bull* **1984,** *32,* 3323.
86. Denniff, P.; Macleod, I.; Whiting, D. A. *J. Chem. Soc. Perkin Trans 1* **1980,** 2637.
87. Denniff, P.; Whiting, D. A. *J. Chem. Soc. Chem. Commun.* **1976,** 711.
88. Nigam, I. C.; Levi, I. *Can. J. Chem.* **1963,** *41,* 1726.
89. Dev, S. *Tetrahedron* **1960,** *8,* 171.
90. Ramaswami, S. K.; Bhattacharyya, S. C. *Tetrahedron* **1962,** *18,* 575.
91. Matthes, H. W. D.: Iuu, B.; Ourisson, G. *Phytochemistry* **1980,** *19,* 2643.
92. Sharma, S. C.; Shukla, Y. N.; Tandon, J. S. *Phytochemistry* **1975,** *14,* 578.
93. Sharma, S. C.; Tandon, J. S.; Uprety, H.; Shukla, Y. N.; Dhar, M. M. *Phytochemistry* **1975,** *14,* 1059.
94. Sharma, S. C.; Tandon, J. S.; Dhar, M. M. *Phytochemistry* **1976,** *15,* 827.
95. Pai, B. R.; Rao, N. N.; Wariyar, N. S. *Indian J. Chem.* **1970,** *8,* 468.
96. Bernhard, R. A.; Wijesekera, R. O. B.; Chichester, C. O. *Phytochemistry* **1971,** *10,* 177.
97. Henley-Smith, P.; Whiting, D. A.; Wood, A. F. *J. Chem. Soc. Perkin Trans. 1* **1980,** 614.
98. Tuntiwachwuttikul, P.; Pancharoen, O.; Reutrakul, V.; Byrne, L. T. *Aust. J. Chem.* **1984,** *37,* 449.
99. Pancharoen, O., M.S. Thesis, Mahidol Univ., Thailand, 1982.

100. For other syntheses of rubranine *see* Bandaranayake, M. W.; Crombie, L.; Whiting, D. A. *Chem. Commun.* **1969**, 58; Kane, V. V.; Grayeck, T. L. *Tetrahedron Lett.* **1971**, 3991.
101. Combes, G.; Vassort, P.; Winternitz, F. *Tetrahdreon* **1970**, *26,* 5981.
102. Nomura, T.; Fukai, T.; Hano, Y.; Uzawa, J. *Heterocycles* **1982**, *17,* 381.
103. Kanjanapothi, D.; Pantong, A., Department of Pharmacology, Chiangmai Univ., unpublished data.
104. Jaipetch, T.; Reutrakul, V.; Tuntiwachwuttikul, P.; Santisuk, T. *Phytochemistry* **1983**, *22,* 625.
105. Tuntiwachwuttikul, P.; Pancharoen, O.; Reutrakul, V., unpublished data.
106. Pancharoen, O.; Patrick, V. A.; Reutrakul, V.; Tuntiwachwuttikul, P.; White, A. H. *Aust. J. Chem.* **1984**, *37,* 221.
107. Lawrence, B. M.; Hogg, J. W.; Terhune, S. J.; Pichitakul, N. *Appl. Sci. Res. Corp. Thailand,* Report No. 2, 1, 1971.
108. Mongkolsuk, S.; Dean, F. M. *J. Chem. Soc.* **1964**, 4654.
109. Dechatiwongse, T.; Yoshihira, K. *Bull. Dep. Med. Sci. Thailand* **1973**, *15,* 1.
110. Tuchinda, M.; Srimaruta, N.; Habanananda, S.; Kanchanapee, P. Dechatiwongse, T. *Siriraj Hosp. Gaz.* **1984**, *36,* 1.
111. Casey, T. E.; Dougan, J.; Matthews, W. S.; Nabney, J. *Trop. Sci.* **1971**, *13,* 199.
112. Amatayakul, T.; Cannon, J. R.; Dampawan, P.; Dechatiwongse, T; Giles, R. G. F.; Huntrakul, C.; Kusamran, K.; Mokkhasmit, M.; Raston, C. L.; Reutrakul, V.; White, A. H. *Aust. J. Chem.* **1979**, *32,* 71.
113. Tuntiwachwuttikul, P.; Pancharoen, O.; Jaipetch, T.; Reutrakul, V. *Phytochemistry* **1981**, *20,* 1164.
114. Tuntiwachwuttikul, P.; Limchawfar, B.; Reutrakul, V.; Pancharoen, O.; Kusamran, K.; Byrne, L. T. *Aust. J. Chem.* **1980**, *33,* 913.
115. Dinter, H.; Hänsel, R.; Pelter, A. *Z. Naturforschung. C* **1979**, *35,* 154, 156.
116. Kuroyanagi, M.; Fukushima, S.; Yoshihira, K.; Natori, S.; Dechatiwongse, T.; Mihashi, K.; Nishi, M.; Hara, S. *Chem. Pharm. Bull* **1980**, *28,* 2948.
117. de Pascual, J.; Bellido, T. I. S.; Gonzalez, J. M.; Muriel, M. R.; Hernanlez, J. M. *Phytochemistry* **1981**, *20,* 2417.
118. Connell, D. W.; Sutherland, M. D. *Aust. J. Chem.* **1969**, *22,* 1033.

0939-1/86/0204$06.00/0

Chapter 14

Alkaloid Components of Zizyphus Plants

BYUNG HOON HAN* and MYUNG HWAN PARK*

The ethnopharmacology of sanjoin, the seeds of *Zizyphus vulgaris* Lemark var. *spinosus* Bunge, is of great interest to modern pharmacological and chemical scientists because sanjoin has been used as a hypnotic agent in Chinese medicine to treat neurologic insomnia. Daechu, the fruit of *Zizyphus jujuba* Miller var. *inermis* Rehder, is another ancient medicine of interest to modern scientists because it has been frequently prescribed by Chinese medical practitioners with unproven results.

Current research on the chemical and pharmacological aspects of sanjoin in oriental and folk medicine includes uses as hypnotics, sedatives, and nervine tonics for insomnia and arrythmia. The following pharmacological aspects of sanjoin are being studied: hypnotic *(1)*, tranquilizer *(2)*, sedative *(3–6)*, papaverine-like *(3)*, analgesic *(3)*, antiinflammatory *(3)*, antiarrhythmic *(7)*, and hypotensive *(8)*. Studies on the chemistry of flavonoids have included swertisin and spinosin *(9, 10)*, 6‴-sinapoylspinosin, 6‴-feruloylspinosin, and *p*-coumaroylspinosin. Betulinic acid *(1)*, betulin *(11)*, jujubogenine, ebelin lactone *(11)*, and jujubosides A and B *(12, 13)* have been studied to understand the chemistry of terpenoid saponins.

*Natural Products Research Institute
Seoul National University 28
Yunkeun-Dong, Chongro-Ku
Seoul 110, Korea

Many scientists showed deep interest in the ethnopharmacological aspects and suggested that the effective compounds are alkaloids *(2)*, but no one could isolate these compounds. Therefore, the role of the alkaloids in sanjoin has been the subject of debate until recently. Saponins and flavonoids were reported in connection with hypnotic activity, but the saponins' ED_{50} values were shown to be as high as 500 mg/kg *(5, 6)*.

The daechu fruit is used to treat chronic bronchitis, consumption, and blood disease and as an analeptic, an expectorant, a sedative, and a taste modulator. Components of the fruit are triterpenoids *(14, 15)*, *Zizyphus* saponins I–III *(16)*, jujuboside B *(16)*, zizybeosides I–III *(17)*, naringin *C*-glycoside *(17)*, cAMP and cGMP *(18, 19)*, and ethyl α-D-fructofuranoside *(20)*. The stem bark contains terpenoids *(21)* and cyclopeptide alkaloids *(22)* (mauritinin A, mucronin D, amphibine H, nummularins A and B, and jubanines A and B). The leaf contains yuziphine, yuzirine, coclaurine, asimilobine, isoboldine, and norisoboldine *(23)*

Some peptide alkaloids were isolated from the stem bark of the daechu tree, but no hypnotic activity was found *(22)*.

We isolated the alkaloidal components and tested the alkaloids for hypnotic activity. The sanjoin alkaloids were named tentatively sanjoinines A, B, C, D, etc., based on their mobility on the chromatogram. Similarly, the daechu alkaloids were named daechu alkaloids A, B, C, etc., and the names daechuines S1, S2, S3, etc., were used for the alkaloids from the stem bark of *Z. jujuba* var. *inermis*.

Table I shows the sedative or hypnotic activity of the extract of sanjoin and its various fractions. As shown in Table I, oral administration of the methanol extract prolonged the hexobarbital-induced sleeping time more than 67% as compared to that of a control animal when both were treated intraperitoneally with 50 mg/kg hexobarbital. The butanol fraction was even more potent, whereas the benzene- and ether-soluble alkaloid fractions showed less activity. From 60 kg of sanjoin we could obtain only 20 g of the alkaloid fraction and 450 g of the butanol fraction.

Table II shows the same activity test with the alkaloid fractions obtained from daechu. Both the ether- and butanol-soluble fractions showed strong activity.

When repeated silica gel flash column chromatography and preparative thin-layer chromatography were combined, several

Table I. Sedative Activity of Sanjoin

Z. vulgaris var. *spinosus* semen (5.5 kg)
↓
benzene extract (1865 g) → alkaloid fraction (0.65 g)
↓
methanol extract (440 g) → diethyl ether fraction (180 g) ↓ alkaloid fraction (0.52 g)
↓
butanol fraction (46 g)
↓
water fraction (213 g)

Fraction	*Sleeping Time (min)*		*Increase in Sleeping Time (%)*
	Control	*Sample*	
Methanol extract (1.0 g/kg)	27.7	46.5	67.8
C_6H_6 alkaloid fraction (50 mg/kg)	23.6	29.2	23.5
Ethanol fraction (0.5 g/kg)	27.7	30.3	9.3
Butanol fraction (0.5 g/kg)	28.8	41.2	43.0
Water fraction (0.5 g/kg)	25.0	28.8	—

NOTE: Sodium hexobarbital (50 mg/kg) was injected intraperitoneally 1 h after the sample was given orally (n = 6).

Table II. Effect of Daechu on Hexobarbital-Induced Sleeping Time in Mice

Fraction	*Control*	*Sample*
Diethyl oxide-alkaloid fraction (10 mg/kg)	14.8	22.1
Butanol fraction (0.3 g/kg)	18.0	28.6
Water fraction (1 g/kg)	18.0	19.0

NOTE: Results are in minutes.

alkaloids were isolated and crystallized. Their structures were elucidated by chemical and spectral analysis (Tables III–V) *(23–26)*. Isolated yields were very low, falling in the 10.3–10.5% range. Sanjoinine A was the most abundant and was the only sample in which enough material was obtained to test for sedative activity (Table VI).

As shown in Table VI, pretreatment of mice by oral administration of 3 mg/kg sanjoinine A prolonged hexobarbital-induced sleeping time 60% or more than that of the control animal. These data suggest that sanjoinine A will show the activity even with a dosage lower than 3 mg/kg.

Sanjoinine A was identified as the known cyclopeptide alkaloid, frangufoline *(27–30)*, by chemical and spectral analysis.

Table III. Alkaloids from Sanjoinin

Compounds	*Molecular Formula*	*Molecular Weight*	*mp (°C)*	*[α]$_D$ (deg)*	*Yield (%)*
Sanjoinine A	$C_{31}H_{42}N_4O_4$	534	249	-316	6×10^{-3}
Sanjoinine B	$C_{30}H_{40}N_4O_4$	520	212–14	—	5.5×10^{-6}
Sanjoinine D	$C_{32}H_{46}N_4O_5$	566	256–58	-53.6	4×10^{-5}
Sanjoinine F	$C_{31}H_{42}N_4O_5$	550	228	-215	1.3×10^{-4}
Sanjoinine G1	$C_{31}H_{44}N_4O_5$	552	236–38	-68.6	3.5×10^{-5}
Sanjoinine G2	$C_{30}H_{42}N_4O_5$	538	182	-79.2	1.6×10^{-4}
Sanjoinenine	$C_{29}H_{35}N_3O_4$	489	281–82	-272.5	2.2×10^{-4}
Sanjoinine E[a]	—	—	166	-146.2	2.7×10^{-5}
Sanjoinine Ia[b]	—	—	155–57	-140	1.2×10^{-4}
Sanjoinine Ib[c]	—	—	184	—	8.7×10^{-5}
Sanjoinine K[d]	—	—	159–161	+35	1.4×10^{-3}
N-Methylasimilobine	—	—	193–95	-204	5×10^{-6}
Caaverine	—	—	204	-80	6.8×10^{-5}
Zizyphusine	$C_{20}H_{24}NO_4$	342	214–16	+317	6.2×10^{-3}

[a]Nuciferine. [b]Nornuciferine. [c]Norisocorydine. [d](+)-Coclaurine.

Table IV. Alkaloids from Daechu

Compound	*mp (°C)*	*Yield (%)*
Daechu alkaloid C (lysicamine)	212	2.4×10^{-4}
Daechu alkaloid E (nornuciferine)	204	5.1×10^{-4}
Daechu cyclopeptide I	114	4.1×10^{-5}
Zizyphusine	214–15	6.2×10^{-4}

Table V. Alkaloids from the Daechu Tree

Compound	*mp (°C)*	*[α]$_D$ (deg)*	*Yield (%)*
Daechuine S1	250	-316	6.3×10^{-5}
Daechuine S2	238–42	-118.6	2.6×10^{-4}
Daechuine S3	192–94	-440	5.9×10^{-4}
Daechuine S4	239–41	-297	3.4×10^{-4}
Daechuine S5	233–35	-421.3	2.8×10^{-5}
Daechuine S6	192	-393.5	6.2×10^{-4}
Daechuine S7	158	-648.3	1.4×10^{-4}
Daechuine S8-1	185–88	-218.2	1.2×10^{-4}
Daechuine S9	115	-481	1.7×10^{-4}
Daechuine S10	126–28	-381.5	6.9×10^{-4}
Daechuine S26	114	—	6.5×10^{-5}
Daechuine S27	226–28	-386	4.8×10^{-4}

Table VI. Effect of Sanjoinine A on Hexobarbital-Induced Sleeping Time in Mice

Experiment	*Control*	*3 mg/kg*	*10 mg/kg*	*30 mg/kg*
1	18.2 ± 6.3	—	—	31 ± 10.4[a]
2	16.3 ± 9.8	26.1 ± 13.1[b]	30.6 ± 19[c]	—

NOTE: Sanjoinine A (3, 10, and 30 mg/kg) was orally administered 1 h before 50 mg/kg sodium hexobarbital was injected intraperitoneally into six mice in each group. Results are in minutes.
[a]Plus 71%. [b]Plus 60%. [c]Plus 87%.

Every peak in the ^{1}H-NMR and ^{13}C-NMR spectra and major peaks in the mass spectrum of sanjoinine A could be reasonably assigned to the given structure of frangufoline. Many of the mass fragment ions of sanjoinine A and other cyclic peptide alkaloids were given structural identity by referring to the work of Tschesche *(28, 29)*.

The a^+ and b^+ fragment ions are the principal and base peaks, respectively, and suggest the size and character of the N-terminal basic amino acid. The j^+ and k^+ ions are produced by deletion of the terminal basic amino acid and ring opening at the ether bond. These ions suggest that the amino acid participates in ring formation. The e^+ (*m/e* 303) and f^+ (*m/e* 190) ions are produced by deletion of the N-terminal amino acid and ring opening at the bond between the two amino acids in the ring. These ions indicate which amino acid is bound to (*p*-hydroxystyryl)amine by the ether bond. These two pairs of ion species are complementary to each other in the elucidation of the ring amino acid sequence. Analysis of the amino acid composition after acid hydrolysis and extensive analysis of the ^{13}C-NMR and ^{1}H-NMR spectra of the unknown cyclic alkaloids supported the structures.

The structures of cyclic peptide alkaloids isolated from sanjoin are shown on page 210. As mentioned previously, sanjoinine A was found to be the known compound frangufoline. The other sanjoinines are believed to be new compounds. All peptide alkaloids except sanjoinine G2 are 14-membered cyclic peptide alkaloids bearing *N*-dimethylphenylalanine as the terminal amino acid.

Sanjoinine G2 is an example of a ring-opened, branched tripeptide bearing a *p*-hydroxybenzaldehyde group. This compound could also be produced from sanjoinine A by ring opening

Structures of Peptide Alkaloids from Sanjoin

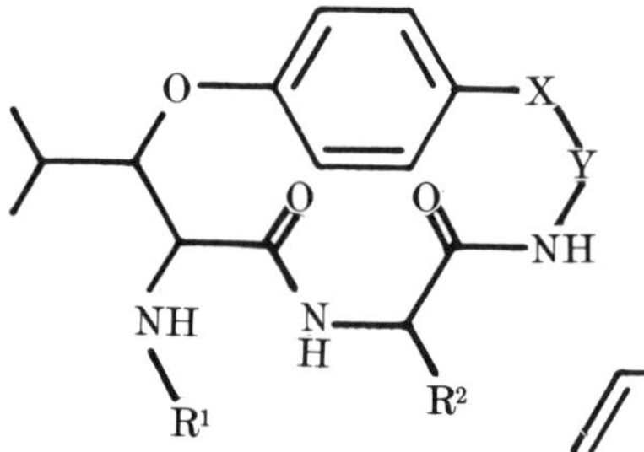

O X Y NH O O NH N H R[1] R[2]

1-6

O O C H O O NH NH N H NH$_2$ O N

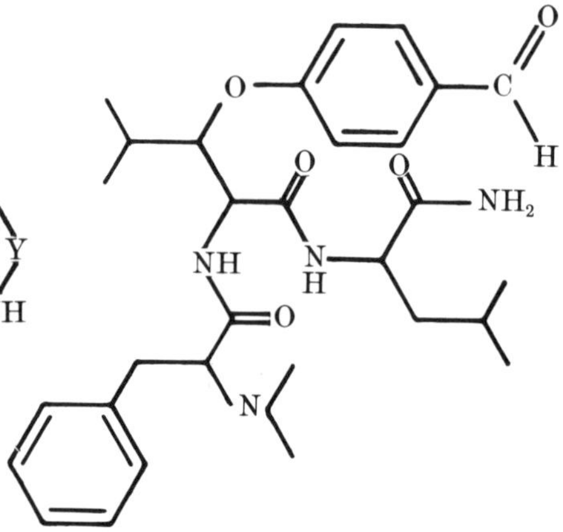

sanjoinine G2

Compound	Number	X-Y	R[1]	R[2]
Sanjoinine A (frangufoline)	1	CH=CH	$(CH_3)_2$Phe-	$-CH_2CH(CH_3)_2$
Sanjoinine B[a]	2	CH=CH	(CH_3)HPhe-	$-CH_2CH(CH_3)_2$
Sanjoinine D[a]	3	$CH(OCH_3)CH_2$	$(CH_3)_2$Phe-	$-CH_2CH(CH_3)_2$
Sanjoinine F[a]	4	CH=CH	$(CH_3)_2$Phe-	$-CH(OH)CH(CH_3)_2$
Sanjoinine G1[a]	5	$CH(OH)CH_2$	$(CH_3)_2$Phe-	$-CH_2CH(CH_3)_2$
Sanjoinenine[a]	6	CH=CH	cinnamoyl-	$-CH_2CH(CH_3)_2$

[a]New.

at the olefinic bond by osmium tetroxide and periodate oxidation. The reaction intermediate sanjoinine dialdehyde was isolated and identified by its ^{1}H-NMR spectra. The reverse reaction consisting of a C-1 unit transfer reaction and cyclization by a ligase is one of the possible biogenetic pathways in cyclic peptide alkaloid formation.

Sanjoinine D could be produced from sanjoinine G1 by diazomethane treatment in the presence of boron trifluoride *(32)*.

The structures of aporphine and tetrahydrobenzylisoquinoline alkaloids isolated from sanjoin are shown on page 211. Zizyphusine is a quaternary base isolated from the butanol-soluble fraction of sanjoin, and zizyphusine is considered at present to be a new compound not described in the literature.

The structures of daechu alkaloids are shown on page 212. Nornuciferene, zizyphusine, oxonuciferene (*see* structure) and a

Structures of Aporphine and Tetrahydroisoquinoline Alkaloids from Sanjoin

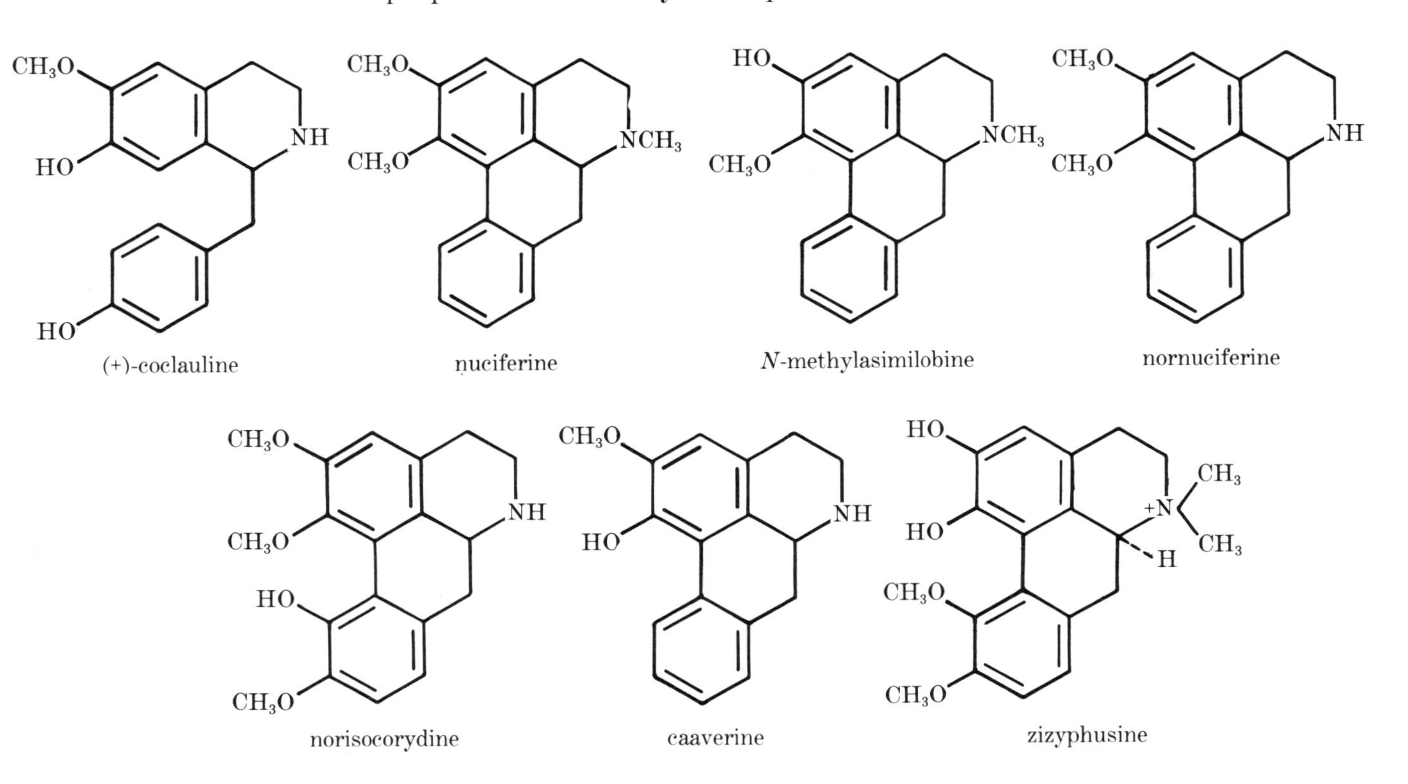

Structures of Alkaloids from Daechu

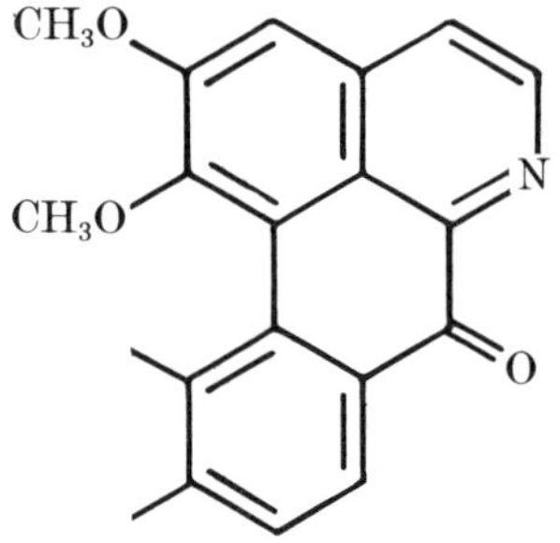

daechu alkaloid C (oxonuciferine)

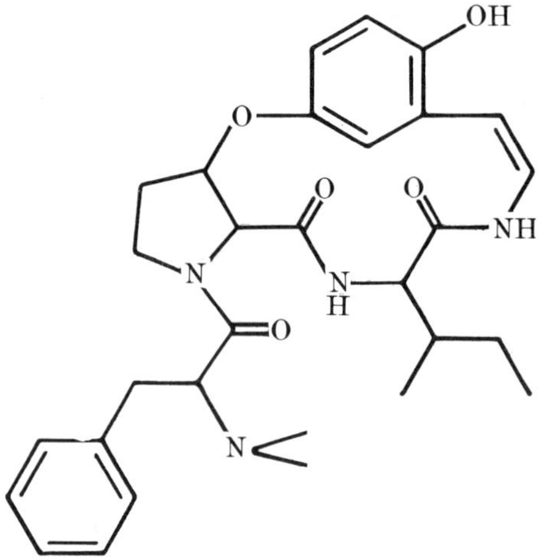

daechu cyclopeptide I

13-membered cyclic peptide alkaloid that was identical with daechuine S26 were isolated from the stem bark of the daechu tree.

On page 213 are structures of the daechuine S*X* series of alkaloids that were isolated from the extract of the stem bark of the daechu tree. The first four alkaloids are 14-membered cyclic peptide alkaloids, and the rest are 13-membered cyclic peptides.

Altogether we isolated 30 alkaloids from the extracts of sanjoin, daechu, and the daechu tree. Among them 20 were peptide alkaloids and 10 were aporphine or benzylisoquinoline

Structures of Cyclopeptide Alkaloids from Stem Bark of *Z. jujuba* var. *inermis*

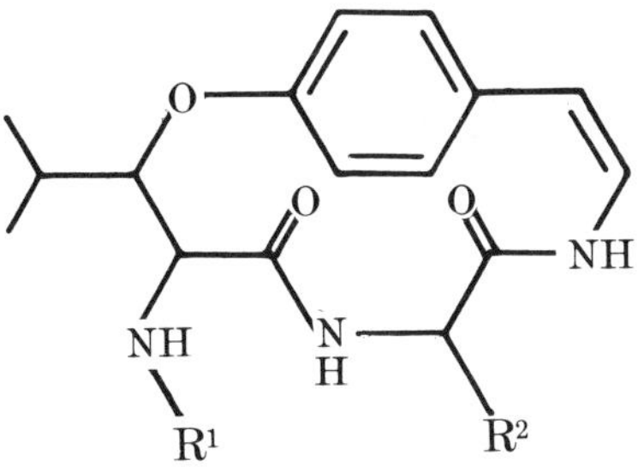

Compound	R^1	R^2
Daechuine S1 (frangufoline)	$N(CH_3)_2$Phe-	$(CH_3)_2CHCH_2$-
Daechuine S2 (frangulanine)	$N(CH_3)_2$Ile-	$(CH_3)_2CHCH_2$-
Daechuine S4 (franganine)	$N(CH_3)_2$Leu-	$(CH_3)_2CHCH_2$-
Daechuine S5 (new)	$N(CH_3)_2$Val-	$(CH_3)_2CHCH_2$-

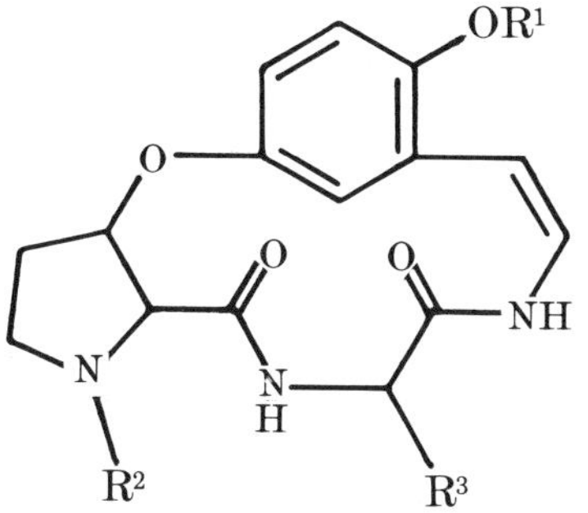

Compound	R^1	R^2	R^3
Daechuine S3 (new)	CH_3	$N(CH_3)_2$Ile-Ile-	$CH_3CH_2CH(CH_3)$-
Daechuine S6 (new)	CH_3	$N(CH_3)_2$Phe-	$CH_3CH_2CH(CH_3)$-
Daechuine S7 (new)	CH_3	$N(CH_3)_2$Leu-	$(CH_3)_2CHCH_2$-
Daechuine S8-1 (new)	CH_3	$N(CH_3)_2$Leu-Leu-	$(CH_3)_2CHCH_2$-
Daechuine S9 (mucronin D)	CH_3	$N(CH_3)_2$Phe-Leu-	$CH_3CH_2CH(CH_3)$-
Daechuine S10 (new)	CH_3	$N(CH_3)_2$Try-	$CH_3CH_2CH(CH_3)$-
Daechuine S26 (new)	H	$N(CH_3)_2$Phe-	$CH_3CH_2CH(CH_3)$-
Daechuine S27 (nummularin B)	CH_3	$NH(CH_3)$Ala-Val-	benzyl

Table VII. Distribution of Alkaloids in *Zizyphus* Plants

Alkaloids	Z. vulgaris *var.* spinosus, *Seed*	Z. jujuba *var.* inermis	
		Fruit	*Stem Bark*
Zizyphusine	+	+	+
14-membered cyclopeptide	+++	–	+
13-membered cyclopeptide	–	+	+++
Aporphine	+	+	–
Tetrahydrobenzylisoquinoline	+	–	–

alkaloids. Of the 20 peptide alkaloids 14 were new compounds not previously reported in the literature.

Table VII is the final summary of the distribution of *Zizyphus* alkaloids in our samples, sanjoin, daechu, and the daechu tree. Sanjoin is rich in 14-membered cyclic peptide alkaloids, whereas the stem bark of the daechu tree is rich in 13-membered cyclic peptide alkaloids. We are now seeking an efficient way of synthesis to get a sufficient amount of the cyclic peptide alkaloids for our structure-activity studies on the neuropharmacological activity.

LITERATURE CITED

1. Kawaguchi, R.; Kim, K. W. *J. Pharm. Soc. Jpn.* **1940,** *60,* 343, 595.
2. Kim, E. C. *J. Pharm. Soc. Korea* **1971,** *15,* 53.
3. Watanabe, I.; Saito, H.; Togaki K. *Jpn. J. Pharmacol.* **1973,** *23,* 563.
4. Shibata, M.; Fukushima, M. *Yakugaku Zasshi* **1975,** *95,* 465.
5. Woo, W. S.; Shin, K. H.; Kang, S. S. *Korean J. Pharmacog.* **1980,** *11,* 141.
6. Shin, K. H.; Woo, W. S.; Lee, C. K. *Korean J. Pharmacog.* **1981,** *12,* 203.
7. Cho, T. S.; Hong, S. S. *Korean J. Pharmacol.* **1976,** *12,* 13.
8. Ahn, Y. S.; Kim, K. H.; Cho, T. S.; Kim, W. J.; Hong, S. S. *Korean J. Pharmacol.* **1982,** *18,* 17.
9. Woo, W. S.; Kang, S. S.; Wagner, H.; Seligmann, O.; Chari, V. M. *Phytochemistry* **1980,** *19,* 2791.
10. Woo, W. S.; Kang, S. S.; Shim, S. H.; Wagner, H.; Chari, V. M; Seligmann, O.; Obermeier, G. *Phytochemistry* **1979,** *18,* 353.
11. Shibata, S.; Nagai, Y.; Tanaka, O.; Doi, O. *Phytochemistry* **1970,** *9,* 677.
12. Otsuka, H.; Akiyama, T.; Kawai, K.; Shibata, S.; Inoue, O.; Ogihara, Y. *Phytochemistry* **1979,** *17,* 1349.
13. Inoue, O.; Ogihara, Y.; Yamasaki, K. J. *Chem. Res., Synop.* **1978,** 144.
14. Yagi, A.; Okamura, N.; Haraguchi, Y.; Noda, K.; Nishioka, I. *Chem. Pharm. Bull.* **1978,** *26,* 1798.
15. Yagi, A.; Okamura, N.; Haraguchi, Y.; Noda, K.; Nishioka, I. *Chem. Pharm. Bull.* **1978,** *26,* 3075.

16. Okamura, N.; Nohora, T.; Yagi, A.; Nishioka, I. *Chem. Pharm. Bull.* **1981,** *29,* 676.
17. Okamura, N.; Yagi, A.; Nishioka, I. *Chem. Pharm. Bull.* **1981,** *29,* 3507.
18. Cyong, J. C.; Hanabusa, K. *Phytochemistry* **1980,** *19,* 2747.
19. Cyong, J. C.; Takahashi, M. *Phytochemistry* **1982,** *21,* 1871.
20. Yagi, A.; Koda, A.; Inagaki, N.; Haraguchi, Y. *Yakugaku Zasshi* **1981,** *101,* 700.
21. Ikram, M.; Tomlinson, H. *Planta Med.* **1976,** *29,* 289.
22. Tschesche, R.; Kohkhar, I.; Wilhelm, H.; Eckhardt, G. *Phytochemistry* **1976,** *15,* 541.
23. Ziyaev, R.; Irgashev, T.; Israilov, I. A.; Abdullaev, N. D.; Yunuev, M. S.; Yunuev, S. Y. *Khim. Prir. Soedin.* **1977,** *2,* 239.
24. Han, B. H.; Park, M. H. *Program 32nd Annu. Conv. Pharm. Soc. Korea* **1983,** pp. 147-48.
25. Han, B. H.; Park, M. H. *Program 33rd Annu. Conv. Pharm. Soc. Korea* **1984,** p. 105.
26. Han, B. H.; Park, M. H. *Program 32nd Annu. Conv. Pharm. Soc. Korea* **1983,** p. 165.
27. Han, B. H.; Park, M. H. *Program 33rd Annu. Conv. Pharm. Soc. Korea* **1984,** pp. 93-94.
28. Tschesche, R.; Last, H. *Tetrahedron Lett.* **1968,** *25,* 2993.
29. Tschesche, R.; Wilhelm, H.; Fehlhaber, H. W. *Tetrahedron Lett.* **1972,** *26,* 2609.
30. Bishay, D. W.; Kowalewski, Z.; Philipson, J. D. *Phytochemistry* **1973,** *12,* 693.
31. Tschesche, R.; Rheingans, J.; Fehlhaber, H. W.; Legler, G. *Chem. Ber.* **1967,** *100,* 3924.

0939-1/86/0215$06.00/0

Author Index

Subject Index

A

G

H

I

J

K

L

M

N

O

P

Copyediting and indexing by Deborah H. Steiner
Production by Meg Marshall

Text and cover design by Pamela Lewis

Managing Editor: Janet S. Dodd

Typeset by Hot Type Ltd., Washington, DC
Printed and bound by Maple Press Company, York, PA